THE NEGLECTED GARDEN:
The Politics and Ecology of Agriculture in Iran

THE NEGLECTED GARDEN:

The Politics and Ecology of Agriculture in Iran

KEITH McLACHLAN

I.B. TAURIS & C° L^{td}
Publishers
London

Published by I.B.Tauris & Co. Ltd.
3 Henrietta Street
Covent Garden
London WC2E 8PW
England

Copyright © 1988 Keith McLachlan

All rights reserved. Except for brief quotations in a review, this book, or any part thereof, must not be reproduced in any form without permission in writing from the publisher.

British Library Cataloguing in Publication Data

McLachlan, K. S.
  The neglected garden : the politics and
ecology of agriculture in Iran.
  1. Food supply—Political aspects—Iran
I. Title
338.1′9′55   HD9016.I72

ISBN 1-85043-045-4

Typeset by Oxford Computer Typesetting
Printed and bound in Great Britain by
Redwood Burn Limited, Trowbridge, Wiltshire

To Anne
whose help in preparing this study was unstinting

Frontispiece: (i) Iran — General.

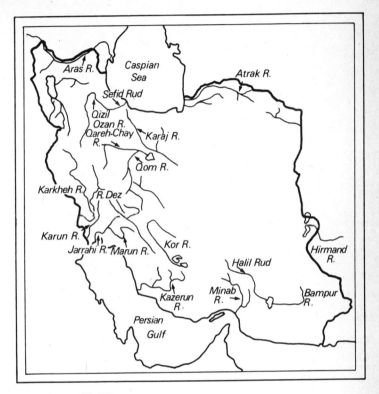

Frontispiece: (ii) River systems.

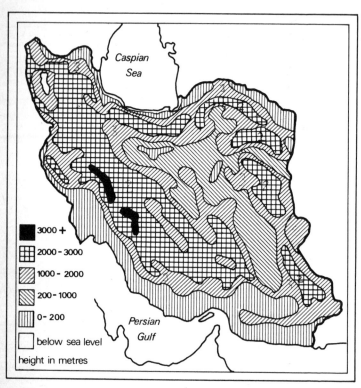

Frontispiece: (iii) Relief.

CONTENTS

Preface	xv
Acknowledgements	xvii
Glossary	xix

1 Introduction: Changing Patterns of Agriculture — 1
  1.1 Introduction — 1
  1.2 The natural resources debate — 4
  1.3 Structural problems and agrarian reform — 6
  1.4 Agriculture and the impact of oil — 8
  1.5 Conclusion — 10

2 The Environment and Agricultural Ecology — 11
  2.1 Environmental constraints — 11
    2.1.1 The limitations of topography and soil — 12
    2.1.2 The constraints of rain and snow fall — 17
  2.2 Changes in the surface area under cultivation — 22
  2.3 Conclusion — 24

3 National Political Evolution and Agriculture in the First Half of the Twentieth Century — 27
  3.1 The constitutional movement and agriculture — 27
  3.2 The rise of the Pahlavi dynasty: Reza Shah and the lowly status of agriculture — 31
  3.3 The fall of Reza Shah: the military occupation of Iran and the failure in agriculture — 40

4 The Context for Change: Agricultural Development and the Move towards Land Reform 1946-63 — 47
  4.1 Agriculture in the years of crisis 1946-54 — 47
  4.2 Monarch and *majles*: a struggle for power — 51
  4.3 Concentration of political control under the shah: the context for agrarian reform — 61

5 Water Resources and Irrigation Cultures — 65
5.1 Introduction — 65
5.2 The development of water laws — 69
5.3 The irrigation regions — 75
  5.3.1 The peripheral basins and plains of the central plateau (Western Kavir basin and the Kavir margins) — 76
  5.3.2 The central desert areas (Dasht-e Kavir and Dasht-e Lut) — 77
  5.3.3 The Mashhad basin, the eastern sumps, the Zagros mountains and the eastern Gulf coast — 79
  5.3.4 The mountain rims of the Alborz and Western Zagros — 81
  5.3.5 The plains of Khuzestan and the northern Gulf coast — 83
  5.3.6 The Caspian plains — 84
  5.3.7 Other irrigation regions — 85
  5.3.8 The dry farming regions of the north-west and Caspian plains — 86
5.4 Traditional irrigation and *qanat* culture — 89
5.5 The survival of the *qanat* — 95
5.6 Modern irrigation — 99
5.7 Conclusions — 102

6 Change and Development in the Countryside: Land Land Reform and Centralization — 105
6.1 The move to land reform — 105
6.2 The land reform — 111
6.3 Farming structures and reform — 121
6.4 The organization of cultivation — 125
6.5 Water as a factor in production — 129
6.6 Agricultural credit — 132
6.7 Reinforcement of commercial farming: agro industry — 134
6.8 The regional effects of reform — 137
6.9 The shah's reform programme and changes in rural society — 142
6.10 Conclusions — 149

7 Agriculture, Oil and Development Planning — 153
7.1 A review of planning — 153
7.2 Planning for agriculture: agriculture in an oil-based economy — 157
7.3 Agriculture in the development plans — 164
7.4 Agriculture and structural change — 178

7.5 Agricultural planning and regional change	183
7.6 Conclusions	187

8 Agriculture, Revolution and the Rural Community — 189
8.1 Agriculture in the prospectus of the revolution	189
8.2 Agriculture in the Five-Year Plan	193
8.3 The administration of agriculture and agricultural development	199
8.4 Land reform, land laws and the effects of insecurity of land ownership	203
8.5 The rural community after the revolution	213
8.6 Conclusions	221

9 Can Agricultural Self-Sufficiency Be Restored? — 225
9.1 Introduction	225
9.2 An assessment of self-sufficiency	230
9.3 The self-sufficiency problem: a balance sheet – the positive factors	234
9.4 The self-sufficiency problem: a balance sheet – the negative factors	242
9.4.1 The problem of rising demand	243
9.4.2 The future of agricultural output	247
9.4.3 Government pricing policies for agricultural goods	250
9.4.4 Iran: the oil economy	254
9.5 Trends in agricultural self-sufficiency: the outlook for the future	260
9.6 Conclusion	264

Notes	267
Index	297

FIGURES

FRONT
    General map   vi
    River systems   vii
    Relief   viii

Chapter 2
2.1 Geographical regions   12
2.2 Rainfall   18

Chapter 5
5.1 Land use in 1974/5 and 1977/8   67
5.2 Irrigation regions   76

Chapter 6
6.1 The contributions of traditional and modern sectors and agricultural output 1975   137
6.2 Regional variations in maximum size of holdings in land purchased from share-croppers under phase two regulations   138
6.3 Regional effects of phase one of land reform to October 1972   139
6.4 Regional distribution of transactions in phase two   141

Chapter 7
7.1 Oil revenues paid to Iran 1954-85   160
7.2 Second plan allocations to agriculture and water resources   166
7.3 Third plan allocations to agriculture and water resources   169
7.4 Fourth plan allocations to agriculture and water resources   171
7.5 Fifth plan allocations to agriculture and water resources   174
7.6 Capital information in agriculture 1963/4 to 1976/7   179
7.7 Trends in agriculture's contribution to GDP   180

Chapter 8
8.1 Value added in the agricultural sub-sectors 1963/4-1976/7   195
8.2 Size of farms 1972/3   211
8.3 Sources of income for rural households   215

Chapter 9
9.1 Estimates of agricultural product self-sufficiency 1967, 1974 and 1984   231

9.2 Iranian agricultural imports 1983-5   233
9.3 Food production and consumption 1972   244
9.4 Crop production 1358-63 (1978/80-1984/5)   248
9.5 Trends in domestic and foreign supplies of rice, sugar and wheat   250
9.6 (i) The financial and economical role of oil   255
    (ii) The impact of oil revenues on food imports and exports   256
9.7 Agriculture in GDP 1978/9-1983/4   260

Preface

Agriculture is among the most important social and economic foundation stones of Iranian life and culture. In this book I am concerned with the causes and effects of transition and change in rural Iran in the twentieth century. During recent decades the countryside has been increasingly exposed to the policies of the central government through the development of communication systems and the government's growing financial power provided by oil revenues.

Rural conditions during the first half of the century were far from idyllic; insecurity prevailed in many areas and few sections of the population had reliable or profitable tenure of their properties. The arrival of development planning and an apparently determined effort to improve the domestic economy in the years following the Second World War offered the farmers the prospect of better water supplies, easier access to urban markets and more generous provision of agricultural credit. The inception of land reform in January 1962 appeared to open the way for a radical alteration in the structure of land ownership for the benefit of the cultivating peasantry and small proprietors. The way in which these opportunities for agrarian change were managed is the main subject of this book. The broader economic context within which the land reform and other agricultural policies were implemented forms an integral part of the study since what I diagnose as often inept and generally disingenuous approaches to the rural economy on the part of the authorities prepared the way for thoroughgoing and mainly adverse developments in agriculture, arising from the expansion of the petroleum industry.

Successive governments' treatment of agriculture and rural

society allowed the considerable skills of Iranian farmers in water and land management to be undervalued; at the same time, the agricultural potentialities were overestimated. This study concludes that there was serious mishandling of traditional farming and a failure of modern agricultural enterprises to replace it. The advent of the Islamic republic brought promises of a new future for agriculture and the village communities.

I visited Iran briefly in 1983 and thus include a review of agricultural performance in the years immediately following the 1979 revolution (Chapter 8). It would be gratifying to suggest that the villages had become prosperous and the outlook for agricultural production had become bright. However, on the basis of personal experience and continuing research into the progress in the Iranian countryside, I remain unconvinced of any major improvement, and agricultural self-sufficiency, as I show in Chapter 9, now appears to be entirely out of reach. A more conclusive verdict must inevitably await access to more reliable and geographically comprehensive data than have been available so far.

Such indicators as are available must be viewed as mere straws in the wind pointing towards a modest degree of growth in a few regions, stagnation in many others and decline in the remainder. Until better evidence, preferably from the field, can be analysed, I adhere to the view that the damage done to water supply, agricultural management, agronomic skills and social structures by events in the period 1963-78 fatally compounded the existing problems of farming. Perhaps the oil wealth was destined to relegate agriculture to a diminishing role in national development, however enlightened the policies of the central government. The authorities and their policies increasingly became hostages to the good financial fortune of the petroleum sector, especially in the period after 1973, though long before then the villages were showing an inability to compete with the rapidly expanding economic opportunities available in the towns. In any event, the attempted agricultural revival during the 1980s failed significantly to halt the decline of the farming industry, and only the most determined optimist can assume that this decline will not persist irreversibly.

Acknowledgements

My thanks are due to many past and present colleagues who have assisted my studies since I first moved to the School of Oriental and African Studies in London. Professor A.K.S. Lambton deserves my particular gratitude, not only for teaching me in her Persian language classes and for giving me an understanding of Iranian agriculture over more than twenty-five years, but also for reading through the manuscript of this book.

I owe a great debt to those Iranians who were unstintingly helpful in village, provincial town and capital. I should like to mention the Alavi, Barghi and Majd families together with Husain Asghari, Susan and Reza Dowlatshahi, Husain Mahdavy, Hasan Mehran, Senobar and Showkat Tafazoli, Reza Sanguri, Reza Soudagar, Kazem Zarnegar and all those other Iranians who contributed to my fieldwork and my education. The staff of the universities of Tehran, Melli (Martyr Beheshti) and Jondishahpur (Ahwaz) were generous in their assistance and I take pleasure in mentioning Ahmad Mostofi, Mehdi Mohaqeq, Ehsan Naraqi, Taqi Razavian and Cyrus Salmanzadeh, who took great pains to enable me to benefit from Iranian literature and research among the rural communities.

There are many American, British and European scholars to whom I am deeply indebted. Most of them will find my acknowledgements in the form of citations of their written words in the text. The British Institute of Persian Studies provided me and my family with a home and a library in Iran and my gratitude is due to Brian Spooner and David Stronach, formerly at the Institute in Tehran. I offer my sincere thanks for their encouragement to my fellow academics at the British Institute, including, among others, Marianne and David Brooks, Michael Burrell, Brian Clark, Bob Rehder, and Andrew Stobbs. Michael Bonine was also good enough to read through a section of this book and offer helpful advice.

I also thank those staff, students and administrators at the School of Oriental and African Studies who aided my efforts, not least Tony Allan, John Bracken, Michael Gatehouse,

Lieut.Col. Moyse-Bartlett, Edith Penrose and Sir Cyril Phillips. Finally it is a great pleasure to record my appreciation to my former tutors and friends at the Department of Geography, University of Durham, especially Howard Bowen-Jones and the late Bill Fisher.

My thanks also to Sue Harrop who designed and drew the maps and diagrams with patience and skill. Iradj Bagherzade and his staff at I.B. Tauris encouraged me in this study and special recognition is due to them for their patience and interest — in particular, to my editor, Mrs Margaret Cornell.

Glossary

ab-ambar	Traditional water storage chamber, usually covered by a dome.
ab-gha'ri	Drowning water
ba'er	Barren land
band-e jim	Note 'C', a provision of the land reform law of 1 March, 1980 which permitted the distribution to peasants and others of land unutilized by land owners.
bandkari	The practice of constructing temporary weirs as water diversion works.
boneh	A team of peasant cultivators jointly providing at least the labour factor in share-cropping cultivation.
Bongah-e Abiari	The Irrigation Institute.
Bonyad-e Mosta'zafin	The Foundation for the Oppressed.
dasht	Plain
daym	Dry land cultivation
enqelab-e sefid	The reforming 'white revolution', introduced by the shah in 1963.
farsakh	Equivalent to approximately six kilometres.
Feda'iyan-e Eslam	A loose organization of Muslims dedicated to the protection of (shi'ite) Islam. Founded in 1946, individuals from the organization were responsible for assassinations of those deemed to be a threat to Islam.
Firqeh-e Demokrat	The Democratic party of Azarbayjan. A formation of mainly former communist party members established in Tabriz in 1945.
gardesh	A cycle within a rotation.
gavband	The owner of oxen or other draught animals who contracts to plough within peasant-landlord share-cropping contracts.

haqq-e abi	Rights to a share of water supply.
haqq-e nasaq	Rights to share in the village ploughlands.
haqq-e risheh	Right to cultivation in the village ploughlands.
haraseh	See *boneh*.
harim	The protected area around a water well within which other wells cannot be sunk.
hava bin	*Qanats* fed from a seasonally fluctuating water table.
Iran Novin	The New Iran party established by Hasan Ali Mansur in 1963. It became the ruling party until the troubles of the late 1970s.
jub	An open irrigation ditch.
juft-e gav	A pair of draught oxen.
jui	An open irrigation channel normally associated with river offtake irrigation systems.
khaleseh	Crown land.
khoshneshin	Villagers without rights to cultivation or permanent employment on the land.
khordeh malek	Small independent proprietor.
Majles	The national assembly.
Majles-e sena	The upper house of the national assembly.
Mardom party	The Peoples party. This grouping formed in 1954 was led in the *majles* by Asadollah Alam.
marja' al-taqlid	The man designated as the most learned and righteous of the religious classes in Shi'ite Islam.
mawat	Literally 'dead' lands that have remained uncultivated.
madar-chah	Mother well or water-yielding feeder well of a *qanat*.
Melliyun party	The National Party, led by a close associate of the shah, Dr Eqbal, established in the 1950s as the official alternative to the Mardom party within the *majles*.
moqanni	The digger of adits and wells that make up the *qanat*.
Mojahedin-e Khalq	A guerrilla organization which was formed in 1965 combining Islamic, nationalist and socialist sentiments in a party which was dedicated to the overthrow of the monarchy.
molla	A cleric of the Shi'ite religious classes.

Glossary

motavalli	Guardian of an institution in receipt of income from an endowment.
nasaq	The pattern of village ploughlands.
'omdeh malek	A large landowner usually holding substantial properties in one village or more.
payab	Subterranean access to a *qanat* used in villages and towns to draw water for household purposes.
pilevar	Itinerant pedlar.
pishkar kani	The process of moving a mother well of a *qanat* further into the yielding portion of the water table.
posht-ab qanat	See *hava bin*
qanat (*qanavat* pl.)	An adit shaft constructed to carry water from subterranean aquifers to a spot on the surface at a variable distance from the water source.
qarq-ab qanat	A *qanat* the mother well of which draws its water supply from the permanent water table.
saddi qanat	*Qanats* drawing water from dams.
sahra	See *boneh*
sar boneh	Leader of a peasant cultivating team.
shabaneh ruz	A period of a day and a night. In respect of a *qanat*, a 24-hour cycle.
tuyul	A land assignment.
'Ulama	Men learned in Islamic teachings.
'urf	Traditional or customary legal practice, not necessarily sanctioned by Islamic law or the provisions of the Civil Code.
vaqf (*awqaf* pl.)	Fixed property endowed to a charity. Usually land, water or buildings.
vaqf-e 'amm	An endowment in perpetuity to the public good.
vaqf-e khas	An endowment to a personal or family charity in perpetuity.
vaqfnameh	A certificate of endowment of property.

1 Introduction: Changing Patterns of Agriculture

1.1 Introduction

During the Anglo-Iranian oil crisis of 1951-3, the Iranian Government took the view that the Iranian economy could survive without oil revenues[1] and reliant on domestic agriculture, while the outside world, on the other hand, could not manage without Iranian oil.[2] These assumptions on the part of the government at that time illustrate the basic misconceptions concerning the Iranian economic system which the analysis presented in this book in many ways refutes. At the present time no less than in the 1950s, the belief prevails that the country has an agricultural strength which enables its economic foundations to be untouched by the effects of the oil industry or indeed the modernizations that were attempted under the two Pahlavi monarchs.

The purpose of this book will be to show that the substructure of Iranian farming and rural society was fundamentally altered during the present century. By the mid 1960s, a mere decade after the oil nationalization, the country no longer had the ability to call on reserves of rural production, labour or skills. Even in 1951-3 the economic crisis was weathered only with privation.[3] Later on, it could be argued that the Iranians had altogether lost the ability to feed themselves, though faith in the concept of the undiminished agricultural heartland has remained widespread.

The causes of change are not difficult to find. The discovery, exploitation and export of crude oil gradually but inexorably led to the transformation of the economy from one mainly sustained by farming and craft manufacture to one externally supported by oil revenues. The importance of agricultural output inevitably declined with the rise in oil production — particularly after the rise in oil revenues in the mid-1950s. That agriculture should stagnate or decline in absolute importance as a provider of national income was always a risk as the effects of oil revenues took their economic and social toll. Oil income, and its role in creating 'rentier' conditions,[4] has been particularly pernicious in its effects on agriculture in Middle East oil-exporting states.[5] But the extent and rapidity of agricultural decline in Iran would also seem to have resulted from inconsistent policies and government negligence.

Neglect — possibly culpable neglect — of the farming population compounded the adverse features brought about by the development of the oil economy. It will be argued in this book that the authorities took for granted the resilience of agriculture in the face of accelerating economic and social change. The defence that they were encouraged by foreign allies and advisers to believe that modernization of the rural sector was politically imperative[6] would appear unconvincing in the light of the deep Iranian conviction that Iranian agriculture can only be understood by Iranians.

A more likely source of official complacency was the conventional attitude that, although agriculture had previously lived through long periods of official negligence, it none the less had the resource base to maintain its central position in the economy.[7] In addition to misplaced confidence in agriculture's ability to accommodate itself to the impact of oil wealth there was also a measure of antagonism towards this backward and intractable sector. The several well-meaning advocates of rapid economic growth at the Ministry of Economy during the 1960s and 1970s found agricultural and provincial development programmes both expensive and slow.[8] Improved rates of production in petroleum, manufacturing industry, transport and services were rarely matched by performance in agriculture. The rates of growth in agriculture in 1963-8 and 1968-73 were 3.4 and 4.4 per cent respectively, whereas the average rate of growth for the economy as a whole was 8.6 and 9.4 per cent for the respective plan periods.[9] Poor prospects for agricultural development made it unattractive to investment, and appeared to lead successive governments to the conclusion that manpower and other resources were best diverted to other sectors[10] — into industry, construction and the country's economic infrastructure.

Once traditional agriculture had become secondary to changes elsewhere in the economy in the mid-1960s it followed that government intervention would be either to establish new farming structures that would be more responsive to government needs or to exploit the countryside for the benefit of the urban population. In a range of actions taken during the 1960s and 1970s — from land reform to food pricing policies — the state sought to manipulate agriculture to its own ends. Interference was capricious, episodic and partial but almost invariably damaging.

Government policies towards the agricultural community, whether occasioned by negligence, incompetence or the belief that traditional modes of production obstructed the country's ability to establish a brave new world on a modern economic base, ignored the essential nature of much of Iranian farming. The present author would contend that, in the face of a hostile physical environment, cultivation has been sustained only by the extremely skilled use of crop strains, agronomic practices and irrigation techniques. All these virtues were embraced within the traditional rural systems which, despite their well-known inadequacies, were invigorated by innovation and regional variants.[11] In the years after 1973 a combination of inappropriate government policies and the overwhelming effects of oil wealth wrought a dramatic change with the depopulation of the countryside.[12] The male labour force in agriculture declined rapidly, in some villages almost to extinction. The very reservoir of skilled labour that could be deployed at low cost and with great intensity in agricultural production was reduced throughout the countryside. Survival of the rural community after so severe a haemorrhage was far from certain, especially in view of the problems of rural disruption faced in the years following the 1979 revolution.

This book will trace the course of events affecting agriculture from the early years of the twentieth century and will examine its treatment under the Pahlavi monarchs, culminating in the land reforms of the 1960s. Water and irrigation are crucial components of Iranian agriculture but their management by the authorities during the transition from traditional to modern farming ignored the pervasive importance of water ownership and control systems in the many villages where the *qanat* system of water supply was in use. Chapter 5 will show how the complex patterns of organization surrounding the *qanat* and the use of other underground water supplies were severely undermined in the modern period. A similar train of events occurred simultaneously in land ownership as traditional relations in land tenure and use were demolished (Chapter 6). In both cases it was found less easy to substitute new structures for the existing ones than had been imagined. The processes of economic planning (Chapter 7), which in many ways served the country well after 1963, failed to come to grips with the intrinsic difficulties of promoting growth in traditional agriculture, which was finally abandoned in favour of imported farm technology and imported foodstuffs.

Following the revolution, a dynamic revival of traditional agriculture was expected. Unfortunately, the continuing problems caused *inter alia* by the authorities' past mismanagement, preoccupation with agrarian reform, and a lack of sustained interest in the welfare of the rural community prevented any appreciable development.

1.2 The natural resources debate

It can be said with some truth that there are two highly polarized views of Iranian agriculture. First, there is the optimistic approach,[13] which suggests that agricultural potential is considerable and may be realized simply by the use of modern methods of farming[14] or by the improvement of those agrarian structures that underpin traditional areas of cultivation. It would appear that most Iranians — and the various official policies[15] — subscribe to this positive view.

The second view is best characterized as one of dispassionate realism. Its exponents have long, but mainly non-Iranian, roots. Lord Curzon,[16] John Murray,[17] Gideon Hadary[18] and Professor A.K.S. Lambton[19] all in their various ways foresaw that attempts to improve Iranian agriculture would encounter problems of confused and insecure land ownership, regional maladministration or political expediency at the centre as well as shortages of fertilizers, credits and irrigation water supplies. It is no accident that this view prevailed before the decade of rapid economic growth, 1965-75. For the most part, it also predated the inception of land reform in the early 1960s. The reason why one must go back in time to discover elements of realism in any assessment of the prospects for Iranian agriculture[20] is probably that those who wrote on this subject after 1962 were preoccupied with the events surrounding implementation of the land reforms. In even more cases, commentators were beguiled by economic progress in the oil, industrial and urban service sectors into believing that either low agricultural production was of little importance or that poor performance in agriculture could be made good once other urgent priorities in petroleum and manufacturing industry had been attended to. It was not until 1975 that serious disquiet was voiced unequivocally on the unsatisfactory contribution of agriculture to national wealth, food supply and rural welfare.[21]

It is perhaps useful to add another perspective to the debate: namely, to examine critically the extent to which real resources of water and land exist and to establish how far they provide adequate bases for the expansion of farming. At the same time, it is relevant to evaluate whether crop yields are improving or not; whether the country is finding more or less food from domestic sources; indeed, whether Iran has not crossed the threshold into permanent dependence on imported foodstuffs and agricultural raw materials. Much of the Iranian dilemma arises from its legacy from the past and, above all, its position as an oil-based economy.[22] None the less, the fact remains that Iran, with a population of some 50 million which could grow to 71 million by the end of the century,[23] has a deteriorating ratio between natural resources available to agriculture and mouths to feed. Equally, industry appears destined to draw most of its agricultural raw materials from outside the country in default of reliable and abundant local produce.[24]

As a by-product of the debate, arguments for and against land reform, which were so popular among the ideologically committed during the 1970s, must be seen as conflicts over technicalities rather than as principles.[25] It might be accepted that the agrarian structure has been appreciably modernized in the period since 1961 and that various ministers have achieved some reclamation of new land for cultivation as, for example, in the Dez project located in the south-western province of Khuzestan.[26] But, after more than ten years of expensive tampering with land tenure and high-cost dam building, it is clear that no enormous potential has been revealed through which production of crops can be augmented.[27] The high degree of constraint on crop production imposed by the physical environment has not been significantly alleviated by these development activities.

1.3 Structural problems and agrarian reform

For centuries agriculture has been depressed by constricting systems of land tenure, onerous taxation and periods of extreme insecurity.[28] Insecurity affected landlords and peasants alike and was exacerbated by factors arising from the geography of the country. Central governments rarely maintained their authority throughout the country on a consistent and well organized basis. For long periods the provinces were beyond the day-to-day reach of the capital, leaving them at the mercy of regional governors, powerful landlords and tribal chieftains, whose treatment of the settled cultivators could be a powerful disincentive to rural peace and productivity.[29]

Problems also arose from the conflicts of nomadic pastoral and sedentary cultivating groups. While there was often a considerable measure of interdependence between the two, banditry, extortion and competition for land or water resources were also common afflictions of areas where nomadic herding and arable farming went on side by side.[30] Lurestan, Kurdestan and parts of the Zagros were the domains of powerful nomadic tribes, and the settled population was frequently disturbed by their activities.

At the same time, the existence of a powerful landlord class apparently did much to suggest that farming was hampered by their extractions from and mistreatment of the peasantry.[31] There were elements of truth in this assessment,[32] though the actual situation in the villages also depended on the individual landlord, the character of his bailiff if the owner was an absentee, and the local practices surrounding crop division, servitudes and rights to cultivation. *Haqq-e risheh*, for example, gave specific rights to peasants to share in the work on the cultivated area of a landlord village.

There was a certain justice in the belief that the problems of Iranian agriculture — maldistribution of land ownership together with maladministration and insecurity of personal rights — were largely man-made. Remedies were long felt to be available such as the land reform which was proposed by the Democratic Party during the Constitutional period, 1906-21. These, it was supposed, could be effected through political action — an assumption that grew in strength during the twentieth century. Yet the inadequacy of legislating a solution for the country's agricultural deficiencies was identified well before inauguration of the land reforms in 1962. Professor Lambton pointed out as early as 1953, 'It is futile to suppose a movement for reform can be brought about by an act of the legislature alone'.[33] Unfortunately, this caveat was not heeded. Policy was to be dominated over twenty-five years by government attempts to solve the problems of slow growth and apparently low productivity by restructuring the sector in the mistaken belief that once proper organization of government control and land ownership had been effected, agriculture would prosper.[34]

The legislative and structural approach brought few rewards. Considerable changes during the shah's regime, often at great human and financial cost, met neither the criteria laid down by the governments of the time nor the food needs of the population at large.[35] The years since 1979 have proved that agrarian reform is no easier for the revolutionary authorities than for the ancien regime. There can be little doubt that some changes will be required if the sector is to meet basic objectives such as greater food self-sufficiency. There must be a clear policy towards land ownership so that this unnecessary form of insecurity can be eliminated. In other countries of the region, notably Iraq and Syria, the failure to distribute sequestrated lands and the continuing annual leasing led to land abandonment and poor levels of cultivation of the remaining areas.[36]

Farmers in Iran would also benefit from pricing policies which minimized the disincentive to produce and favoured a longer-term view of investment in farm assets, whether orchards or machinery. Management of the wheat market in the years 1983-6 under a government purchasing programme illustrated how influential pricing policies could be.[37] Gradual improvement nation-wide in credit facilities, marketing services, storage capacity and provision of farm inputs such as fertilizers and seeds could enable considerable gains to be made in farm efficiency. Such suggestions are not an academic matter. In 1984, the Persian-language magazine for farmers, *Keshavarz*, listed ten problems affecting agriculture, most of which could be redressed without need for any type of land reform. Better budgetary management, improved import controls, direction of credits to long-term investment and better co-ordination of government intervention were key elements.[38]

1.4 Agriculture and the impact of oil

Government agencies' preoccupation with structural reform precluded proper weight being given to the environmental difficulties faced by the farming community, whether subsistence, collective or capitalist. Shortages of water for irrigation, land of indifferent fertility, lack of labour, persistent crop diseases and isolation from the main markets remained as constraints for which the state had offered few palliatives. Yet in operational terms these factors were equally, if not more, important to most farmers than the issues of farm size and changes in tenurial conditions that so exercised the authorities and ideologues of all kinds in Tehran.

Belief in the improvements to be had from land reform tended to disguise a second important negative pressure on agriculture — the changes occurring in other sectors of the economy. In fact, there is a strong case to be made for the argument that in the pre-oil economy agriculture retained importance as an employer, a creator of domestic wealth and an earner of foreign exchange only because alternative means of achieving the same objectives were absent. The growth of the Iranian oil industry from the first decade of the twentieth century, and its impact on the national economy, was underestimated as a force for change. It was implicitly assumed during the 1960s that land reform or other forms of agricultural development had only to produce a relative improvement in rural prosperity as compared with the pre-reform situation.[39] This was a gross oversight which became increasingly difficult to justify with the passage of time. Rising urban wage rates, growing employment opportunities outside agriculture and an addiction to a more sophisticated life than the village could provide left farming in most areas of the country as a poor option for all but the immobile. Traditional ways of life and employment in the countryside could not be expected to survive *in toto* the effects of modernization. Some degree of change was inevitable. But the pace and scale of modernization, forced by the growth in oil revenues after 1973, affected most aspects of rural life. A number of farming areas benefited through new investment, mechanization, extended water supplies and access to profitable urban markets.[40] But these districts were few, and the poorer areas suffered a great decline in fortunes relative to the towns. In many villages visited by the author in the period 1971-8, even those who had gained considerably from land reform and other government policies were induced, no less than their poorer neighbours, to move out of agriculture by the attraction of better wages elsewhere.

In many ways the departure of so many from the farming community, estimated at between two and three million in the period 1966-76,[41] is a testimony to the truism that poor rewards from agriculture were tolerated only while no alternative existed. Those same poor returns, it is argued, were above all a function of a physical environment that was inhospitable to agriculture, which could be practised only by the deployment of enormous amounts of skilled family manpower which was unpaid. This study will demonstrate that Iranian farming is best likened to gardening. Gains are made painfully and patiently from gardens built on poor soils in an arid climate. The skills involved in coaxing more production from a difficult physical environment are considerable and for the most part extremely labour-intensive. Iranian agriculture has in a short period of time suffered major losses of its skilled farmers which will not be made good immediately even in the most encouraging of political or economic climates. The danger is that the losses will continue and with them the decline in the traditional village cultivated lands.

1.5 Conclusion

Given the constraints of the natural environment, only a false optimism will lead to the survival of attitudes which rest on the belief that miracles of modern technology working within government-controlled management structures will achieve a restoration of the country's agricultural prosperity. Certainly, images of land reform and of the indestructability of the great Persian garden appeared to dominate official policies towards agriculture long into the post-revolutionary period. The lessons of the immediate past and the importance of the underlying environmental realities that affect farming are, it would seem, slow to be learned.

2 The Environment and Agricultural Ecology

2.1 Environmental constraints

A vital background to any discussion of the changing pattern of agricultural land use is a review of the main environmental factors which inhibit production on existing farms and deter reclamation of new areas for cultivation. In truth, geography has not been altogether kind to Iran.[1] On the one hand, the country is blessed with a great diversity of physical features but, on the other hand, many of these, such as high mountain deserts and inland drainage sumps, are of no direct agricultural use.

Conventional interpretations of the landscape suggest no less than eighteen main regions and many more sub-regions[2] (Figure 2.1). Contrasts are immediately apparent. In the centre lies the vast triangular plateau of the Persian heartland. Although the Dasht-e Kavir and Dasht-e Lut deserts dominate the central region,[3] the surrounding basins, for example those of Qazvin-Karaj and Esfahan, represent some of the most fertile agricultural land. The largest area of intensive cultivation lies in the northern borderlands, especially in the Caspian lowlands and parts of northern Khorasan. The high mountain systems of the Alborz and the Zagros have obvious importance in attracting rainfall and creating a multiplicity of local microclimates, but otherwise add only modest areas to the sum of the nation's agricultural domains. In the north-western region, normally regarded as a rich agricultural province, very broken topography reveals a considerable variety of slope and soils

Figure 2.1. Geographical regions.

types which, chronically affected by damaging forms of erosion, give a low average performance from an agricultural point of view.

2.1.1 *The limitations of topography and soil*

A detailed analysis of the topographic and ecological composition of the country shows how powerfully the underlying geology and soil endowment affect the situation. In the north the Alborz runs on an alignment from west to east as a continuation of the European Alpine structures. The western and southern rim of the great central plateau is made up by the

Zagros mountain chain, which again shows a continuity with the Dinaric Alps. In the southern Zagros especially, the rocks have been strongly folded, faulted and overthrust along axes running roughly northwest to southeast. Within the central plateau lies the ancient massif against which the Alpine upheavals bore, though even this central block is ruptured and surrounded in close proximity by largely unstable structures.

The Iranian plateau constitutes a vast area of diverse geographical characteristics. Its centre is taken up by two major deserts, the Dasht-e Kavir lying to the north and the Dasht-e Lut lying to the south of a low mountain zone which runs from Khorasan in the east to Ardestan in the west. These two great deserts are themselves made up of a series of smaller discrete areas and are broken up by mountain ridges, some of altitudes greater than 1,500 metres. Much of the Dasht-e Kavir is covered with surface deposits of salty clay, salt crust and brackish ooze flows. Dasht-e Lut presents an even harsher environment. It comprises an axial basin enclosed between the mountain chains of Kerman and Qa'en-Birjand, with its northern sector slightly elevated and the southern area, known as Kavir-e Zangi Ahmad, somewhat lower and dominated by *namakzar* (or salt deposits). The Lut also contains the *shahr-e lut*, small areas heavily eroded by the action of the wind and looking like abandoned villages. To add to the difficulties of human occupation, there is a broad belt of dunes spreading down its eastern flank.

The extremely inhospitable environment of the deserts permits only scattered and limited areas of cultivation. Underground sweet water is scanty and rainfall negligible. There are a number of oases throughout the region which illustrate the ingenuity of Iranian agriculture in making good use of poor resources. Shahdad, for example, is a palm-growing centre in the Dasht-e Lut, where date output is small but of high quality. Other oases lie on the routeways through the deserts and traditionally provided for a limited number of inhabitants and for the needs of caravan traffic.

If the great deserts of Dasht-e Lut and Dasht-e Kavir provide some of the most highly developed instances of Iranian agricultural management of a difficult environment using sophisticated water provision and horticultural systems,[4] it is the rich basins and plains that lie around them that make up the main farming areas of the country. In Esfahan, Yazd, Kerman and other smaller centres there is a conjunction of widespread if not abundant underground water, moderately reliable rainfall and extensive areas of fertile soils. A prosperous agriculture developed in these areas permitting significant exports of grain, fruit and vegetables.

In the western districts, the plateau merges with the mountain systems of the Zagros. The higher and colder regions of Azarbayjan attract heavy and reliable rainfall as do adjacent parts of the periphery of the plateau such as Hamadan and Kurdestan. Rainfall, often as snow, in the north-west and west is over 300mm per year, permitting dryland grains and temperate fruit crops to be grown. Natural pastures in the higher areas of the west provide a basis for a comparatively dense livestock population kept by village people and semi-nomadic tribal groups. Irrigation is practised in most parts of the north-west using water drawn from rivers, lakes and streams as well as from subterranean sources.

The rim of the central plateau in its south-eastern quadrant makes up much of the province of Baluchestan. Here the surface topography is extremely broken and at altitudes of over 1,000 metres, with volcanic peaks such as Kuh-e Taftan characteristic features of the area. The only district of agricultural importance lies in the Khash plateau, where there is a combination of rich fertile soils and readily extracted underground water. In the Iranian Mekran the landscape is even harsher than in the more westerly parts of Baluchestan, with highly disturbed geological formations giving rise to a complex and generally forbidding hillscape. The Makran coastal strip is lower in altitude than the rest of the region and benefits from both the presence of underground water and the penetration of monsoon rains. Despite these advantages, however, its isolation and a history of insecurity have meant that its agricultural development is poor.

The Jaz Murian depression in south-east Iran remains an internal drainage basin characterized by salt flats and swamps. On the higher slopes of the area the scanty population lives by herding and some settled agriculture as in the Bampur to Iranshahr zone. The Jaz Murian has potential for agricultural development and the Jiroft region attracted considerable investment during the 1970s.

The eastern rim of the plateau, the East Persian Highlands, is aligned on a north-south axis between the city of Mashhad and the mountains of Baluchestan. The topography in the region is varied. High peaks such as the Kuh-e Sorkh at more than 3,000 metres alternate with low cols such as the Gonabad area and lowlands in the region of Torbat-e Haydarieh. For much of the eastern rim the soil cover is patchy but often fertile. Rainfall is low and not always reliable but a profitable dryland grain culture is pursued throughout much of the area, especially in the districts north of Gonabad. Underground water resources are abundant but frequently at considerable depth and this is an area of extremely long *qanats* (adit shafts), sometimes reputed to be more than twenty kilometres from mother well to surface outlet. The eastern villages suffer from the effects of the powerful 'wind of a hundred and twenty days', which sweeps persistently through the area and can cause significant crop losses.

The borderlands with Afghanistan in Sistan are made up of large depressions to the east of the main Eastern Highlands. The Hamun-e Hirmand and Hamun-e Sabari are lakes fed by the Hirmand river system running in from Afghanistan surrounded by large areas of level fertile land. Formerly known as the bread basket of Iran, the Sistan area now carries only a modest population cultivating grains, kirkubitz and cotton.

The Alborz, one of Iran's two great mountain ranges, is geologically and topographically spectacular, rising to 5,654 metres in Mount Damavand and averaging some 3,100 metres throughout its length. Rainfall in the western and central Alborz range averages over 600mm and there is a considerable forest cover remaining on its upper slopes. From an agricultural point of view, however, the region is one of small immediate interest. Narrow valleys and scarcity of low altitude basins mean that cultivation is precarious and has been confined to tiny areas. The less well watered eastern extension of the Alborz running between Gorgan and Mashhad offers a more developed agriculture. In the broad valleys and foothill areas of the River Atrak in the west and the Kashaf Rud in the eastern sector there is a fertile if mainly thin soil cover which supports irrigated cultivation and dryland grains, together with a large livestock population. Large oases such as Mashhad, Nayshapur and Sabzevar make use of the wide basins and plains that occur in the lower sections of the eastern Alborz.

The Zagros mountains run at altitudes averaging over 3,000 metres in a continuous line from south east to north west along the western rim of the central plateau (Figure 2.1). They attract a heavy and reliable rainfall but, like the Alborz, have never provided a basis for a thriving settled agriculture. Soils are poor throughout much of the western Zagros and the climate is one of extremes. Only Shiraz has developed as an important area of permanent agricultural development, much of the rest of the range being occupied by tribal nomadic pastoral groups. In the eastern Zagros altitudes are lower on average, and the existence of wide river valleys and intermontane basins has encouraged some restricted areas of permanent cultivation as well as pastoral nomadism. Nayriz, Fasa and Estebanat cultivate grains, fruits and sugar beet. The narrow and poorly watered coastal plain along the Persian Gulf coast supports a handful of small village communities with very low standards of living except on the coastal plain around Bandar Abbas, where a combination of fertile soils and provision of irrigation water makes for rich though small-scale commercial vegetable growing.

In total contrast to the rest of the country, the great plains of Khuzestan in south west Iran comprise a large and continuous lowland. The area is an extension of the Mesopotamian region. It is a zone of thick alluvial soils drained by major river systems running down from the Zagros to the Shatt al-Arab or the Persian Gulf. Despite the availability of water and fertile soils and a history of intensive cultivation in the distant past, the Khuzestan plains are agriculturally of little contemporary national significance, apart from a cluster of rich irrigated gardens around the River Karun producing early vegetables and fruits, some settled and contract cultivation of grains or other irrigated crops on the higher plain and shifting cultivation with herding outside the irrigated zone. Beginning in the 1950s the government attempted an integrated development of the region through the construction of a series of dams on the main rivers. Provision of irrigation waters to newly reclaimed lands at Haft Tappeh and Dezful indicated that the deep silts had potential for the cultivation of sugar cane, fodders, grains and early soft fruit. It was also clear, however, that the arid environment, featuring sustained high temperatures, hot sand-laden winds from the south and occasional floods, was extremely fragile and of use to farming only with the exercise of great care.[5] None the less, the plain of Khuzestan is one of the few areas of Iran where physical conditions suggest some potential for future agricultural development.

A second great lowland lies in the north of the country, running from Astara in the west to the Turkoman Sahra in the east, for the most part in a narrow shelf of rich, grey peat soils no more than 3 or 4 km wide between the Alborz Mountains and the Caspian Sea. In the east the plain broadens out in the Dasht-e Gorgan where soils are deep and fertile of the podzol type. The region benefits both from a heavy rainfall, everywhere above 600mm though in most places in the west more than 1,000mm, and the flow of water from the north-facing slopes of the Alborz. Water for irrigation uses traditional dams, modern reservoir structures and *qanats*. The warm subtropical climate permits extensive rice growing in the west around Rasht with the lower foothills dedicated to tea, tobacco and fruit, including citrus. To the east more recently reclaimed lands around Gorgan and the Turkoman Sahra are used for cotton monoculture and grain cultivation, while the traditional village lands remain under cotton, grain, vegetables and fruit. The Caspian region was a source of agricultural exports to Russia in the second half of the nineteenth and the early twentieth century,[6] but after that time became an important source of domestic foodstuffs and an exporter of cotton mainly to markets in the Western industrialized countries.

2.1.2 *The constraints of rain and snow fall*

Nothing is so critical to farming in Iran as rainfall. Natural precipitation, whether taken directly into the cultivating cycle or transmitted into subterranean aquifers, is of prime concern. Everywhere, apart from the northern Caspian coastal zone and parts of the north-western region, rainfall is inadequate to support cultivation in most years. More or less half the surface area of the country receives a rainfall of less than 300mm (12ins). For working purposes, this approximates to the area in which dryland cultivation of cereals is marginal or impossible. Some 40 per cent of the country experiences an average annual fall of rain, including snow, of between 300 and 600mm (12-24ins) (see Figure 2.2). Even here precipitation is far from reliable. Great variation affects both the amounts of falls and their periodicity. Distribution of natural precipitation through-

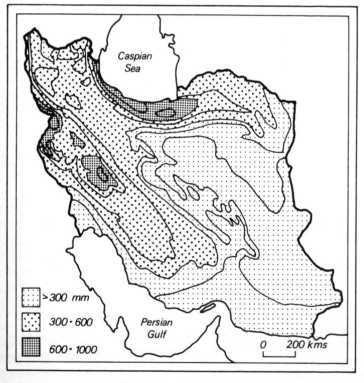

Figure 2.2. Rainfall.

out the year varies enormously from region to region. Sanandaj in the west demonstrates that even in the apparently well-watered Zagros Mountains summer and early autumn are very dry.

Unfortunately, nowhere is there an overall favourable balance in water supply. Seasonal surpluses exist in the Zagros and Alborz Mountains thanks to moderate temperatures and high rainfall conditions. In the remaining 80 per cent of the country[8] the mean water surplus of less than 100mm (4ins) requires irrigation to be practised if agriculture is to survive.[9] Any serious agricultural expansion, therefore, has two principal paths forward. Firstly, it could spread on to the extensive dryland margin comprising those portions of the 20 per cent of the surface area not yet used for settled agriculture but with a rainfall of more than 300mm. Secondly, there is scope for reclamation of new lands under irrigated cropping from the larger zone that is arid in the technical sense.[10]

The conversion of more land to agricultural use in the rainfed areas is possible, though this might be made up of difficult slope and thin soil areas since most of the best and medium lands are already under various forms of cultivation. It would be unreasonable to expect that there could be significant additions to the stock of agricultural land in the rainfed areas of Azarbayjan and the Caspian coastlands in the short term, even in the most favourable economic conditions.

Controversy has always raged concerning the potential for extending the area under irrigation.[11] Those who take an optimistic view appear to believe that the country's rich resources have been under-utilized.[12] A typical stance was that proposed with various modifications by the Plan Organization,[13] indicating that only approximately 20 per cent of all available water is drawn off, and that more than half is lost in evaporation and evapotranspiration from vegetation on pastures and farmlands, the balance being shared in the ratio 1:18 between urban and rural agricultural use. These Plan Organization figures assume that there is a permanent equilibrium in the balance between recharge and extraction of water from underground aquifers. Such an assumption must be regarded with suspicion since extraction rates by pumping have increased greatly over the years since 1963, with little consideration for the maintenance of subterranean aquifer levels. A natural balance was automatically kept by the use of *qanats*, a virtue that disappeared as *qanats* lost favour against use of diesel or electric pumping sets for water lifting.

Efforts were made in the period from 1955 to improve the proportion of water actually converted to use within the country and thereby to reduce losses. Attention was concentrated on saving the outflows from the two great river systems of the Sefid Rud in the north and the Karun with its tributaries in the south.[14] Together with smaller schemes for water storage and flood control, the major new hydraulic works on the Sefid Rud and Karun systems were expected to add as much as 3,000-5,000 million cubic metres of water to the 38,000 million already exploited. There was a remarkably high degree of success under the Third (1962-8)[15] and Fourth (1968-73)[16] Plan programmes for water resource development. In aggregate these dam schemes added more than 10,000 million cubic metres of water storage capacity, not all of which was fully used. Theoretically, there was potential for irrigating up to 760,000 hectares of new or existing land. Some electricity generating capacity was also installed at the sites of the new storage dams.[17]

Despite the admirable progress made in developing irrigation potential in the decade 1963-73 in seeming fulfilment of the aspirations of the optimists, expansion of land under irrigated cultivation went ahead very slowly.[18] Even in the golden years of irrigation development the gains were limited to 145,000 hectares of new land during the Third Plan and 400,000 hectares in the Fourth Plan periods.[19] The following ten years of the abortive Fifth (1973-8) and Sixth (1978-83) Plans represented an almost totally missed opportunity for an expansion of irrigation. Estimates suggest that a mere 220,000 hectares of new land were provided with irrigation water in those years.[20] It should be noted, however, that a measure of improvement was effected on existing areas of irrigation, possibly amounting to 740,000 hectares over the twenty-year span ending in 1983.

The expansion in irrigated farmland was expensively purchased. Financial allocations to water developments amounted to some $1,000 million in the ten years to 1973.[21] These were the years before the oil boom, when finance for development was in short supply, and for a considerable period the demands of the water resources budget left other sectors bereft of both funding and access to real resources such as skilled labour and transport facilities. There was, therefore, a high opportunity cost for the great dam projects implemented in the 1960s.[22]

The government's dedication to hydraulic engineering during the course of the Second and Third Plan periods showed itself only slowly in terms of the area under irrigated cropping. Gains often had a high price, as in the case of the developments in Khuzestan under the Dez dam, where an apparent reclamation of 101,762 hectares, of which 10,000 hectares were used at the Haft Tappeh sugar cane plantation, included only 38,650 hectares of new land under cultivation, less than one third of the total claimed.[23] There was also a great problem in matching operations to the objectives planned by government organizations. At Jiroft near Kerman, for example, it had been planned to reclaim 110,000 hectares using water supplied from sub-surface reservoirs. In the event, less than one tenth of this area was achieved. At Dasht-e Moghan in the north-west of the country there was another instance of an apparently exciting irrigation project that was ultimately to provide only a small addition to cultivated land. High social costs were experienced at the site together with a small net economic gain, since pasturelands were lost to what was often low-grade cultivation on government-sponsored reclamation areas.[24]

The changes that came about in the irrigated area may be summed up — certainly from a crop yield and agricultural production point of view — by looking not simply at the aggregate additions to the irrigated zone, but the quality of the new lands brought into use. In 1950 it was estimated that only 750,000 hectares were given adequate water supplies to be kept under perennial cultivation.[25] Another approximately 750,000 hectares were served by moderate water supply that guaranteed less than full irrigation. The bulk of the *abi* (or watered land), amounting to some two million hectares, in fact had enough water to support supplementary irrigation at best. From those statistics and surveys that are available, whether official or private, it would seem that there was only the most modest of improvements in the quality of irrigation in the twenty-year period ending in 1980. The largest advance was in the surface area of fully irrigated land, where the total doubled to 1,500,000 hectares by 1980. Supplementary irrigation areas increased by only a quarter to approximately 2 million hectares over the same period.[26]

It is important to observe that, among the many changes affecting agricultural land and water use, there was an increasing move on the central plateau from reliance on *qanat* sources to utilization of deep and semi-deep wells in which water lifting was by diesel or electric pumps (see Chapter 5). Estimates of the comparative contributions of *qanats* and deep wells vary enormously. Price[27] suggested that *qanat* systems produced 9,000 million cubic metres of water and wells 8,000 million cubic metres by the mid-1970s. Evidence from the field suggests that the *qanat* has lost further ground since that time. During a field visit to Iran by the author in 1983, for example, official sources showed the *qanats* servicing 500,000-600,000 hectares against the 1,000,000 hectares that they had serviced a decade previously. This situation arose partly as the *qanats* went out of use but also as they were less efficiently maintained or lost their source reservoirs through pumping from adjacent tube wells. The 1983-8 five-year plan included promises of a revival of *qanat* water systems. Financial credits were allocated to improvement programmes for *qanats* under the annual state budgets, though the Ministry of Agriculture appeared for most of these years to lean more towards well drilling than repair of the *qanats* on the grounds that wells were more efficient.

2.2 Changes in the surface area under cultivation

A long-standing trend in agriculture was for abandonment of land in the high mountain areas, especially by settled cultivators in the Alborz Mountains.[28] This process has also affected the more distant and ill-favoured locations of eastern, southeastern and southern Iran,[29] where the struggle against a hostile environment was tolerated only for as long as better economic opportunities did not exist. As Lord Curzon put it, the

Persian peasant was willing to turn to account the scanty resources of nature,[30] but we also know that his perseverance eroded rapidly during the second half of the twentieth century and that Curzon's for once almost flattering descriptions of Iran became less applicable after land reform had begun. In addition to altering the economic balance, the unsettling effects of the first phase of the land reform and subsequent legislation which persisted until the 1970s resulted in the uprooting of the peasantry and the landlord classes, albeit in unequal measure.

The actual surface area of land abandoned is not known. Estimates of the land under cultivation made by different authorities vary considerably.[31] Following the revolution there was a tendency to put the best possible interpretation on trends in agriculture, as in the case of the activities of the Bonyad-e Mosta'zafin (Foundation for the Oppressed).[32] Other organizations had their own special view of the statistics. Examination of figures for agricultural production and of debates in the *majles* suggests that land abandonment *per se* was not seen as the major problem. Rather, discussions at the time suggested that only small areas of physically marginal land had been totally lost to cultivation but that the intensity of land use had deteriorated in general.[33]

One indicator of this change was a fall in the numbers of livestock as shepherds in both the settled and the nomadic sectors became difficult to find. The livestock holding within the country is estimated to have fallen during the 1980s.[34] An extension of rainfed wheat and barley cultivation, a significant proportion of which was on existing farmland, also had the effect of increasing the area of fallow within the grain-based rotation. This trend was established long before the revolution as the area under fallow rose from some 8 million hectares in 1971 to between 9 and 10 million hectares in the mid-1980s.

To summarize the land and water use situation, it may be said that the overall position changed little in the twenty years 1960-80. Some qualitative improvements were made in the irrigated sector, but there were offsetting losses through the less concentrated use of land for farming as rural populations left agriculture and, in many cases, the countryside in increasing numbers throughout the late 1960s and 1970s. Of the total area of Iran, normally given as 164.8 million hectares, 17.5 million hectares, or 11 per cent,[35] was in cropland. This figure was very close to that used by Overseas Consultants Inc. in 1950 and was only slightly under the one in currency with the Ministry of Agriculture in 1984.[36] Within the cropland category there appear to have been fluctuations during the period 1955-85. Only approximately 30 per cent was actually under the plough or tree crops in 1950; by 1970 this was officially claimed by the more optimistic commentators[37] to have reached more than 50 per cent. The agricultural vicissitudes were considerable after 1970 and, even taking into account the rapid expansion of dryland wheat cultivation in the post-revolutionary period, it seems unlikely that this proportion has been improved upon. Indeed, judging by the complaints being made against the Ministry of Agriculture in the *majles*,[38] it was probable that the cultivated area had fallen to below the 1970 level.

Despite the unsatisfactory nature of the statistics, it is clear that during the thirty years of the development of dams, irrigation schemes and reclamation projects there was no perceptible and dramatic expansion on to virgin land nor a measurable improvement in the use of existing cultivated land.

2.3 Conclusion

Iran has a large surface area (1,648,000 square kilometres) by regional and even international standards. The combination of geology, climate and soils has the effect of making some three-quarters of the country of little value to agriculture, herding or forestry. Indeed, as a result principally of environmental constraints, less than 10 per cent is under cultivation of any kind in any one year. At the same time, the broad regions of the central plateau, the mountain rims and the peripheral lowlands have great internal diversity arising from a variety of altitude, slope, soil and water endowments.

Agriculture faithfully reflects the underlying limitations of the physical landscape but, as we shall see later, has elicited a wide range of technical, cultural and agronomic responses with which to reduce or turn to advantage the environmental constraints. Nevertheless, despite this skill and adaptability, the zones of intensive, high productivity farming are comparatively small and, even when aggregated with medium-grade irrigated areas, comprise a mere 3 per cent of the total surface area of the country once forests are excluded. The distribution of agricultural activities, including agriculture-related industries, reveals a regional concentration of farming and processing of agricultural products clearly aligned towards the climatically favoured regions of the north and west.

3 National Political Evolution and Agriculture in the First Half of the Twentieth Century

3.1 The constitutional movement and agriculture

At the end of the nineteenth century Persia was an agricultural country. It is estimated that eight out of every ten Persians were resident in the countryside or dependent upon the land for their livelihoods. Lord Curzon described agriculture at the time as backward and depressed.[1] Certainly, the administration of the country's main source of wealth and the Treasury's principal access to revenues was poor and deteriorating.[2] The combination of an inefficient, capricious and often oppressive tax system, the concentration of local power in the hands of large landowners or tribal khans, the neglect of the extensive crown lands and the absence of disinterested government involvement in the welfare of farmers and farming meant that agriculture was in a depressed state in many areas. Low levels of output and a lack of innovation were its main characteristics. At the same time, drought[3] and other natural rural hazards inhibited progress.

The first change, though it was more an alteration in the theory of land holding than in the practice of agriculture, came with the constitutional movement which flowered in the early years of the twentieth century.[4] The *iqta* system, under which land had been assigned to powerful individuals together with rights to collect revenues in place of the government, was abolished. The holders of land assignments (*tuyul*) had entrenched themselves as leaders and organizers of the villages and accumulated considerable powers over the properties and populations within the areas of their assignments, including a range of irksome duties and servitudes from their peasant communities.

At the same time, the fixed conversion rates (*tas'ir*) at which taxes in kind on agriculture were converted to cash equivalents were abandoned. The rates had for long been devalued as the real prices for cereal crops had risen but the conversion rate had remained unchanged. In theory, the elimination of the *tas'ir* enabled the government to increase its income from taxes on land and lease-holders. In practice, under-reporting, government inefficiency and corruption reduced the revenue expected from the reform.

This watershed in the established system of landholding and taxation marked by the grant of the Persian Constitution in 1906 and the abolition of the *tuyul* did not immediately affect the lives of the peasantry since changes within the *iqta* system had been experienced in some rural areas several years before the constitution was adopted and a number of features of the agrarian system that characterized the pre-constitutional period lingered on long after 1906.[5] Landlords, for example, held on to their land and maintained a strong influence over affairs in their villages, including the use of some peasant servitudes, until the coming of the land reform in the 1960s. But the change in attitudes towards agrarian structure and the growing recognition in the *majles* and among the country's increasingly enlightened intelligentsia of the need for reform of the landlord system made the constitutional movement the beginning of a new phase in Persian economic and social development.

Historians reviewing the contemporary history of Persia place much emphasis on the emergence in 1906 of a constitution which recognized the existence of an executive, legislature and judiciary and laid down, in broad form, the responsibilities of each.[6] Yet the constitutional movement and the promulgation of the constitution have for the most part been an oblique influence on the practice of government during the twentieth century and were less immediate in their impact on agriculture than changes in the legal structure of land tenure would seem to indicate, mainly because the traditional landlord system continued to exist in most Persian villages.[7] The constitution, although never effectively implemented, derives its importance *inter alia* as a rallying point for the forces that have opposed the real managers of the country's fortunes since 1906. At times of stress and perceived weakness in the central authority its ghost has been raised in order to denounce past actions of the government and claim a role in power for those in opposition. During the political crisis of 1978 the religious leaders, the parties comprising the National Front, and the groups of the extreme political left all made reference to the 1906 Constitution in order to validate their complaints against the monarchy and their demands for reform or revolution.[8] But during the years of strong control at the centre it remained primarily as a guideline to be ignored or circumvented as circumstances dictated.[9]

In its early years, the *majles* began the elaboration of important legislation governing individual rights, taxation, and the nature of land ownership. Unfortunately, legislative activity was not matched by the inception of an efficient administration and the imposition of the writ of the legislators across the country. The new shah, Mohammad Ali, who succeeded to the throne in early January 1907, was more hostile to the constitution than his predecessor, but was powerless to prevent completion of the national assembly's legislative work which culminated in October 1907 with the promulgation of the Supplementary Fundamental Law, laying down the duties and rights of the shah, the assembly and the population. The assembly was cautious in imposing new taxation for running the machinery of state and prescribed that economies on the part of the Court would obviate the need for taxes, while other expenses of administration could be covered by domestic loans raised through the new Iranian banks.

Political confrontation between the shah and those supporting the constitution dominated the period 1907-9. The *majles* was closed by the shah in June 1908. However, the continuing agitation by the nationalists in Tabriz, the march on Tehran by Bakhtiari tribesmen, and the movement against the shah in the Caspian provinces, together with interventions by Russia and Great Britain, led to the fall of the shah in July 1909.

The successes of the constitutionalists failed, however, to provide either a sense of unity or motivation to put the affairs of state in order. The reforming zeal of successive but intermittent national assemblies had begun to lose some of its sharpness as early as 1911 largely as a result of Russian pressure. From 1914 to 1921 the major concern was with the depredations of war, the effects of the Russian revolution and the growing influence of Great Britain within Iranian economic and political life. The British attempt to establish what appeared to be a political hold on Iran through the Anglo-Persian Agreement of August 1919 was rebuffed by the *majles* which refused to ratify the arrangement.[10] Political preoccupations and the venting of nationalistic fervour against foreign interference tended to be self-perpetuating as successive governments ignored the problems of reforming the domestic administrative structure and establishing a fair and efficient tax system which alone would have made independence possible. So weak was the authority of the central government that the British had virtual control in the south of the country, the Caspian littoral was in revolt under Kuchik Khan and his partisans, and there were few local notables who were unable to impose their wishes in all important matters, including agricultural affairs, at the expense of the central government.[11]

A further failing of the reform movement was the fragmentation of interests and parties which promoted it. Tribal chiefs seeking their own ends,[12] a religious establishment with conservative leanings, and an odd assortment of radicals pressing for modernization made unhappy political bedfellows. Added to this, the constitutional movement petered out in the political quick-sands of foreign intervention and growing internal chaos. Yet it did achieve some improvements in law and taxation which affected agriculture and rural life. The reform of the system of landholding, and particularly the abolition of the *tuyul* or land assignment, may be seen as a symbol of movement by Iranian agriculture away from its medieval roots and towards an albeit gradual and incomplete modernization some years later.[13]

3.2 The rise of the Pahlavi dynasty: Reza Shah and the lowly status of agriculture

If the constitutional movement led Persia to a legal threshold in 1906 and 1907, the rise of the Pahlavi dynasty following Reza Khan's *coup* in 1921 signalled an even clearer change in society and economy. Where the later Qajars and the *majles* had largely failed to implement their reforms, the first of the Pahlavis, Reza Khan, to an extent succeeded, even though he was far more of a modernizer than a political reformer in the ideological sense that applied to the radicals in the *majles*. Reza Khan was able to impose his writ because he recognized, in a way that his immediate predecessors had never done, the need for a strong army and a centralized administration if changes were to be effected in a country where disaffection from the central government was endemic in the highly varied ethnic, linguistic, and economic regions that together made up the nation state.

The condition of Persia in the period after the First World War was woeful. Arbitrary outside intervention during the war resulted in a measure of economic hardship, especially since a prolonged drought adversely affected agriculture during the same period. Modernization of the country was begun before Reza Khan's assumption of the monarchy. As early as the summer of 1922 the Persian Government sought the assistance of a team of American financial advisers under the leadership of Dr A.C. Millspaugh, whose main task was re-ordering the antiquated and inefficient tax system with special reference to agriculture as a source of Treasury income, a task partially begun earlier by another American adviser, Morgan Shuster. In supporting financial and adminstrative reforms, Reza Khan was motivated above all by the need to strengthen and underpin the capacity of his army. Throughout the Pahlavi period it was a recurring theme that modernization, reform, and improving living standards in agriculture and other sectors of the economy were designed in the first place to serve the immediate political or the longer-term dynastic ambitions of the monarchs and were rarely incorporated into government policies for ideological or idealistic reasons.

Whatever the motivation, Reza Shah's pragmatic approach proved effective in laying the legal and administrative foundations on which later reform could be attempted. The Millspaugh mission found the country in financial chaos, a situation exacerbated by out-of-date records of land ownership and taxation,[14] and put in hand new and improved financial measures which slowly increased the equity, regional comprehensiveness and yield of the tax system. Most taxes related to agricultural land and production, though official registers of villages were inadequate and contained information that rarely conformed to current conditions. The *majles* itself set in motion the long and complicated task of land registration under a law of 1922, and a cadastral survey of the country was begun in 1926 but was completed only in limited and easily accessible areas. During Reza Shah's rule no profound changes were made in the country's financial structure. The Millspaugh mission and the legislative programme of the *majles* achieved a tidying up of the financial system but there was almost no attempt at far-reaching reform.

These conclusions apply particularly to agriculture, where changes in the structure of land ownership and tenure were entered upon only slowly and without conviction. Reza Shah appeared unwilling to offend the major landowners as a group by implementing plans for re-distribution of agricultural lands as proposed by politicians during the constitutional period.[15] The main exception to the government's policy of minimal interference in agricultural affairs was the distribution of public domain lands or *Khaleseh*[16] in 1932 and 1933, when the *majles* authorized transfer of such lands to cultivators in Lurestan, Kermanshah, and Dasht-e Moghan. Little progress was made under this law, however, or under a 1937 law providing for public domain in Sistan to be distributed among the peasants. In part, the abortive nature of the modest essays in land reform during the reign of Reza Shah reflected his own lack of conviction about root-and-branch change, except where the interests of the army were at stake. Also, the strength of the administration, while slowly growing, was altogether unequal to the task of supervising radical change in so basic an area as agriculture, and the government was possibly wiser than its foreign advisers in deciding to leave over the difficult question of ownership and tenure of agricultural lands.

Even modifications to taxation were modest. The revised land tax law, promulgated in 1926, proposed a 3 per cent levy on the product of land, but this was never effective since its implementation relied on a rapid completion of the cadastral survey, which proved impossible. In 1934 the government abandoned the land tax for a combination of income tax and a 3 per cent levy on agricultural produce, including livestock, to be collected on entry to an urban area or on export. The tax load on rural areas was not onerous, though the point is well made[17] that the peasants and landowners were much more constrained by the activities of government officials than had ever formerly been the case, since Reza Shah's centralization policics required a steadily growing bureaucracy, few of whom were directly concerned with the welfare of the farming community.

While reform of the tax structure for agriculture failed to ensure a flow of funds to the Treasury adequate to support the expanding administration and modernization of the army, Reza Shah's government was not slow to see the advantages of state-controlled monopoly trading companies. Direct state monopolies or monopoly trading companies were set up from the early 1930s to manage a number of important commodities, most of them agricultural, including wheat, cotton, dried fruits, gum tragacanth, opium, silk, sugar, tobacco and tea. In addition to providing revenue for the Treasury, the monopolies tended to stabilize prices for agricultural products, provided a guaranteed market and led to some gradual improvement in the quality of agricultural goods. At the same time, Reza Shah's political and strategic requirement of greater domestic self-sufficiency led to the introduction of crops which could substitute for imports. Tea and sugar were two prime commodities affected. Sugar beet cultivation was adopted across a wide area of the country as a result of government encouragement, while considerable areas in the Caspian lowlands were put down to tea plantations with government support. Iran never quite became wholly self-sufficient in either commodity, but domestic production made real savings in foreign exchange and broadened the range of crops and cultivation techniques.

A further significant contribution to agricultural prosperity came as a by-product of Reza Shah's policies in the 1930s for the settlement and control of the nomadic pastoralist tribes. Villages engaged in arable farming had for a long period been at the mercy of raiding tribesmen especially in the Zagros Mountains and the bordering territories, as central authority had diminished towards the end of the Qajar era. Indeed, there is evidence in some areas of a 'beduinization' of large areas formerly under cultivation.[18] Severe measures against the nomads sharply reduced pressure on the settled population and increased security enabled an expansion of arable farming and orcharding, in some regions, such as that around Shiraz, on a considerable scale. Many large nomadic and semi-nomadic groups were subjected to enforced settlement and themselves became increasingly involved in farming. But there were costs attached to the sedentarization policies pursued by the government. The nomadic pastoralists were often treated inhumanely in so far as their adjustment to settled life was given little positive assistance, and many large groups found themselves constrained within physical or climatic areas which were quite incapable of sustaining them in a sedentary role.[19] As a result of the government's policies the livestock holding of the nomadic sector began a slow decline which was to culminate in its virtual disappearance as a significant supplier to the market in meat and livestock products by the late 1970s. Whereas some of the grazing lands were effectively converted to arable farming and orchards, large areas of difficult terrain, often at considerable altitudes, ceased to be used for productive agriculture once the pastoral nomads had withdrawn and there is evidence that the general ceiling on cultivation fell with the termination of cereal growing in upper tribal territory. Few areas above 2,200 metres (7,000 feet) in altitude were under cereals by 1978-9.[20]

Reza Shah provided security for the rural people with one hand, but he also created a degree of uncertainty through his considerable appetite for personal land acquisition. Legally the estates accumulated at that time in the shah's name were purchased, though it is generally recognized that a degree of coercion was involved. Large areas of the northern province of Mazandaran, together with substantial lands elsewhere, were taken in the shah's name before being handed over to the state at his overthrow in 1941. The eventful history of Reza Shah's land acquisitions after that date will be returned to later. The point to be made here is simply that the monarch encouraged the landowners' feelings of insecurity which had chronically depressed the state of agriculture, and at the same time, by becoming a major landowner in his own right, perpetuated traditional landholding and cultivation systems which were clearly inappropriate even then.[21] Insecurity on the part of landowners tended to discourage investment and also to foster absenteeism since the propertied class sought to retrench its position through employment in the upper echelons of the bureaucracy in the capital or in the main provincial cities.[22]

If Reza Shah's interventions into the difficult areas of finance and agriculture were to have only superficial effects, the same cannot be said of his modernization of the legal system and its implications for the position of one of the main bulwarks of Iranian society, the 'ulama or religious classes. It is claimed with considerable justification that the 'ulama suffered as a class, from the reforms of Reza Shah.[23] The creation of state schools was an early blow to the interests of the religious establishment, and harsher measures were to follow when a penal code was adopted in 1926 and a civil code in 1928, both based on Western, mainly French, models. These separated the religious authorities to a considerable extent from administration and execution of the law, other than personal law, much of which had formerly fallen within their purview as administrators of Islamic and customary law. As Reza Shah sought to stimulate the growth of a modern economy, reforms were carried out in the field of commercial law, which was made the sole prerogative of the civil courts from 1932, thereby undermining the traditional role of the 'ulama within the main stronghold of commerce, the Tehran bazaar. A measure that removed most of the remaining links between the 'ulama and the legal establishment came with the promulgation of regulations demanding that all judges be graduates of Iranian or foreign university law faculties.

The large-scale opportunities for employment that had existed for clerics in the pre-reform era were thus much reduced. The income of the religious authorities arising from their monopoly position as legal arbiters and authenticators of commercial documents was greatly diminished. At the same time, new laws were introduced concerning the use of *vaqf* revenues. It is possible that the number of clerics declined during the 1930s.[24] The *'ulama* did not forget the damage done to their reputation, wealth, and influence by the first of the Pahlavis. Though they were unable to make any effective protest against Reza Shah while he was on the throne, their harboured grudges were to play a significant part in grievances aimed against his successor when he, too, turned to modernization programmes at a later date.

Reza Shah's concern with the economy was considerable and during the decade 1931-41 he began the gradual industrialization of the country. It is easy to underestimate the extent of his success in creating a small but useful industrial base.[25] He was beginning industrialization virtually from scratch in a state where all the elements required for modern industry were lacking — roads, railways, electric power, secure water supplies, trained management and skilled manpower. Lord Curzon summed up the situation in Iran in the period before Reza Shah's reign as follows:[26]

> Persia is neither powerful, nor spontaneously progressive, nor patriotic. Her agriculture is bad, her resources unexplored, her trade ill-developed, her government corrupt, her army a cypher. The impediments that exist to a policy of reform, or even to material recuperation, are neither few nor insignificant. ... The outward evidences of decay are numerous and pathetic.

Reza Shah's essay in industrialization must be counted as courageous. The number of factories even at the end of his reign was modest but the expansion of industrial plant was important for agriculture since most factories were based on the use of agricultural raw materials. Processing plants for agricultural produce such as tea, silk, cotton, sugar and vegetable oil made valuable contributions in promoting changes in farming areas. Some provinces, such as those in the Caspian lowlands, enjoyed a measure of prosperity as a result of the introduction of new crops associated with the industrialization programmes. Also, given the limited amount of industrial investment elsewhere in the Middle East at the time and the low base level from which the shah began, it can fairly be claimed that Reza Shah provided Iran with a decade of valuable initiation into the problems of industrialization and set in motion a process that most of his contemporaries thought impossible in the Iranian economic environment of the time.

The 1930s also witnessed an Iranian government tackling the problems of transport and communications in an area where previously nothing had been done or facilities had been set up by foreign companies with their own, often external, interests in mind.[27] The shah himself put great emphasis on the Trans-Iranian railway, which was to link the Persian Gulf port of Bandar Shahpur with Tehran and, ultimately, with the Caspian port of Bandar Shah. The 1,360-kilometre railway remains a monument to his modernizing zeal, though of equal importance is the augmentation both before and during Reza Shah's reign of the all-weather roads, comprising some 3,250 kilometres of strategic highway laid earlier by the British and Russians, by a further 30,000 kilometres of passable, if not paved, road. Improvements in transport brought some immediate and perceptible economic benefits,[28] though the Trans-Iranian railway itself made only a small impact on agricultural marketing despite its high costs.[29]

Nationalistic reasons have been seen as of principal importance in the motivations of Reza Shah's railway construction programme and no small degree of nationalist sentiment was similarly at work in the creation of a national bank (the Bank-e Melli) in 1927, which took over powers of note issue in 1931 from the British-established Imperial Bank. The Banke-e Melli fulfilled Persian aspirations for independence from foreign powers in banking and prepared the way, albeit incompletely, for later expansion of the banking sector as a basis for stimulating and funding development activity. Other important banking institutions which were set up in the 1930s included the

Agricultural and Industrial Bank of Iran in 1933 and the Mortgage Bank of Iran in 1939, though the Bank-e Melli was also involved in expanding the services of the Pawn Bank, first established in 1926, and the National Savings Fund, which was inaugurated in 1939.

While Reza Shah was struggling to bring order to the country at large and making his first tentative essays in economic development, significant changes were being wrought at a much faster pace in the south-west of the country where the Iranian oil industry was emerging as a political and economic enclave[30] quite detached from the mainstream of events in Tehran. Reza Shah was assiduous in reminding Iranians and foreigners in Khuzestan that Iranian sovereignty prevailed: a major crisis between Iran and the Anglo-Persian Oil Company developed in 1932-3 which brought a number of significant victories for the Persian ruler, but which still left Khuzestan's oilfield area as a distinct economic and social region.[31] The oil industry provided Reza Shah and his government with a useful income in foreign exchange. It is estimated that oil receipts accounted for 10 per cent of government revenues in the mid-1920s and as much as 30 per cent between 1929 and 1932, but declined to some 25 per cent in the years 1934-8.[32] At the same time, almost a third of all imports were paid for by oil revenues during the 1930s, as oil eclipsed agriculture as the principal source of foreign exchange and Treasury income.

Reza Shah's achievements have been variously represented as those of an inspired reformer[33] or of a ruler increasingly 'morose and brutal'.[34] He included both these attributes in his complex personality, though at heart he remained primarily concerned with his personal supremacy within the state and with the strength of the armed forces, in which lay his ultimate claim to authority. His attitude to the constitution and the *majles* was coloured by his experience of the years 1906-21, when government was weak and divided, when foreign powers, notably Great Britain and Russia, could intervene almost at will in Persian affairs, and when the country 'seemed to be on the verge of collapse, about to disintegrate into a number of separate parts, some of which might be absorbed into neighbouring states'.[35] He never considered the *majles* to be capable of participating in serious affairs of state, and he quite cynically manipulated or ignored the assembly as the occasion demanded. Iran remained a constitutional monarchy in name, but the position of the shah within the constitution was so powerful as to relegate the elected assembly to comparative unimportance even though, significantly, it was not disbanded. What is more, the reputation of Reza Shah as a strong and unforgiving ruler and his control of the army effectively precluded the rise of any politicians who might have challenged his role within the constitution.

The modernization of Persia can be seen as an extension of the monarch's concern with the army and military affairs. By far the greatest allocation of resources to economic development was to transport and communications,[36] an area of direct military significance since it was roads, and later the railway, that enabled the outlying provinces to be brought firmly within the authority of the capital and permitted the army better access to troublesome regions. Even in the industrialization programme, the objectives were either directly military, such as the armaments factory at Tehran, or were designed to make Iran less dependent on the outside world for basic supplies of foodstuffs and clothing for strategic reasons. The most telling factor in analysing the policies of Reza Shah is his attitude towards agriculture. This sector was above and beyond all else the underlying source of national and personal income and the dominant area of employment, but attracted only passing attention from his government. Changes that were made were largely concerned with taxation. The structure of land ownership and conditions of tenure were left in an obviously poor state and it was not until 1937 that an agricultural development law was proposed, though it was never implemented. A truly reformist government would have been deeply committed to changes in agriculture, but Reza Shah himself became a large landowner of the traditional kind and the vestigial medievalism of Iranian agriculture remained to trouble the governments that succeeded him.

3.3 The fall of Reza Shah: the military occupation and the failure in agriculture

The late 1930s witnessed the consolidation of the personal wealth of the monarch, the further strengthening of the armed forces and a first attempt to initiate the construction of heavy industry. Reza Shah was unchallenged in his absolutist rule, but a small number of intellectuals were attracted by communist ideology and the economic model exemplified in the Soviet Union, and many more Iranians were impressed by the growing might of Germany and the military nationalism of the Nazi party. Reza Shah himself found much to admire in the new Germany, where dictatorship, supremacy of the army, and industrial efficiency were allied to politics directly threatening the imperial powers of Great Britain and Russia. Germany appeared to be the ideal ally for the politically and economically emergent Iranian state. Towards the end of the 1930s, Reza Shah moved into closer relations with Germany, which provided military training and began the construction of an iron foundry near Karaj. At the same time, Germany set up a shipping connection with northern Iran and also fostered a civil airline service between Berlin and Tehran, with a link through to Kabul. There were many large Iranian towns with resident German advisers on technical projects, banking, or military affairs, while the scale of Irano-German trade outstripped all other foreign commercial transactions, apart from oil. The German economic penetration of Iran and the undisguised official and private sympathies of many Iranians with the Nazi regime, while entirely understandable in historical terms, proved to be an unfortunate combination as the world moved towards total war in the period up to 1939 and Iran became caught up, almost inevitably, in its ramifications.

British influence declined rapidly after Reza Shah's accession to the throne, though most Iranians continued to believe that it was the British embassy that controlled events inside the country. There was in fact a paradoxical element in the British position. Reza Shah sought to reduce British involvement in commerce and government and in this he had considerable success. As early as 1921 the situation was described by the

British Ambassador, Sir Denis Wright, as follows: 'By the end of 1921 not a single British adviser remained in the service of the Persian Government: British influence and prestige had reached their nadir.'[37] The commercial and legal capitulations granted by the Qajar monarchs to Britain and Russia were cancelled in 1927 and replaced, in the case of Britain, by a new tariff agreement and jointly approved arrangements for the treatment of British nationals working in Iran. The eclipse of the Imperial Bank as the country's note-issuing agency and the remarkable growth of trade relations between Iran and Germany during the 1930s further emphasized the assertion of Iranian nationalism at the expense of British influence. But the demise of British influence in one area was more than compensated for by its rise in importance in Iran's steadily expanding oil industry. British commercial interests in Iran, far from declining under Reza Shah, became more profitable, of greater strategic significance, and, because of the large state-controlled stake in the Anglo-Iranian Oil Company, of more specific immediate concern to the British Government.

Iran was caught up in the Second World War as a result of the German invasion of the Soviet Union in June 1941. It was seen as the ideal transit route for the flow of supplies to the Soviet armies. The British wish for a secure position in Iran was reinforced by fears that the German forces might break through the Russian lines in the Caucasus and directly threaten the security of the Caspian and Persian Gulf oil fields. A British demand for a reduction in the number of German nationals inside Iran was rejected by the Iranian authorities. On 25 August 1941 Britain and the Soviet Union delivered notes to the Iranian authorities accusing the shah's government of breaching its neutrality through failing to meet this demand and informing the Tehran government that Iran would be occupied by the Allies to ensure a diminution in German influence and to secure a supply route between the Persian Gulf and the USSR.

Following the British and Soviet invasion, Reza Shah abdicated on 16 September 1941 in favour of his son. For the Iranians, the Allied occupation was a deep psychological setback and political affront, evoking memories of the earlier occupation by the great powers that revived strong feelings of resentment against both occupying powers. Dislike was particularly focused on the British from whom a less insensitive response had seemed likely: little better was expected from the Russian Government.[38] The position was made worse by the poor economic state of the country during the war years. Foodstuffs were in short supply since priority was given to the provision of materials to the Soviet Union. The Middle East Supply Centre, operating from Cairo, had limited resources at its disposal and lacked the means of bringing about rapid improvements in Iranian agricultural production. Many of the food supply problems during the 1941-5 period are explained by the activities of the large landowners and the merchants. Hoarding and manipulation of the shortages for profit undoubtably occurred but blame must also be attributed to the USSR, which exported all available cereals from the northern area under its control,[39] and to the failure of Reza Shah's administration, which had done so little to improve the productivity of the farming population.

It is estimated that the price of food trebled between 1941 and 1945, while the general cost of living rose on average by three and a half times over the same period. Iranian dissatisfaction was widespread as inflation raged unabated[40] and the benefits of growing employment opportunities with the military authorities were dissipated by their inflationary effects in a situation of worsening shortages.

The Russians were committed with the British to withdrawal of their troops within six months of the formal cessation of hostilities. But while the Western Allies were withdrawing their forces towards the end of 1945, the USSR was actually reinforcing its position. A revolt by the Pishevari group in November led to the establishment of the Autonomous Republic of Azarbayjan in December, a move immediately emulated in the Mahabad area by the Kurdish People's Republic. It appeared that much of north-western Iran was about to become a permanent zone of Soviet influence. The Iranian Government, alarmed by the consolidation of the Russian position at a time when the Western powers were demobilizing their armies and were deeply disinclined to engender a new round of confrontation with their erstwhile ally, sought assistance from the United Nations. This body urged direct negotiations between Iran and the USSR rather than condemning Soviet intervention in Iranian domestic affairs, and a tentative agreement was reached in March 1946.

The deeply disturbed period of the Second World War and its immediate aftermath was one of economic decline for Iran. Agricultural output remained low and adversely affected by poor weather conditions. Reports from the central and eastern provinces suggest that production fell below local requirements and that starvation levels prevailed for extended periods.[41] The agricultural commodities developed under Reza Shah and the factories associated with them suffered badly from the loss of strong management, the unavailability of spare parts for machinery and, in the northern provinces, the gearing of supply to Russian war needs.

Stimulated in part by wartime shortages of foodstuffs and by an exclusion from a number of other activities by the occupying powers, the Iranian Government made some legislative attempt in 1944 to redress the chronic agricultural problems. The first area of activity was management of the personal estates left by Reza Shah.[42] At the same time, *Khaleseh* comprising long-standing crown lands which extended to some 812 villages[43] were put under the care and maintenance of a state-controlled institute. In practice, the institute did little other than make a start on the registration of land titles. New legislation in 1946 opened the way for the later transfer for sale of crown lands to peasant cultivators. The disposal of *Khaleseh* in Mazandaran[44] and Sistan[45] was marred by inefficiency and corruption on a serious scale and did little to improve the agrarian structure. In Mazandaran, however, the transfer of large areas of land into the hands of a few landowners gave scope for rapid mechanization and the introduction of new crops in some, if not all, estates.

A second piece of legislation which can be seen as the initial step in the modern period towards rationalization of the irrigation sector brought into being an Independent Irrigation Institute (Bongah-e Abiari) in 1943. The ideals behind the legislation were well-founded. Irrigation systems were to be brought under state control. The Institute was founded so that technical skills could be developed, data relating to water availability and water use recorded, and encouragement given to more widespread scientific exploitation of water in agriculture. It had few immediate achievements, though it did provide the basis for a gradual improvement in information on water resources and to an extent made possible the emphasis on dams and reservoirs that developed later in the second and third plan periods (1955-68).

3.4 Conclusion

Iran began and ended the first half of the twentieth century with its economy based on agriculture and its population made up of generally small rural communities. Reza Shah had not left the rural areas untouched despite his apparent lack of a comprehensive and enthusiastic programme for agriculture. His two greatest gifts to them were political stability and crop innovation. Belief that law and order would persist enabled the farmers, both large and small, to invest in their agricultural holdings with greater confidence than they had done for some years previously. Undertakings which took many years of expenditure and risk, such as the digging of *qanats* and the planting of orchards, but which were essential if agriculture was to prosper, were actively pursued. Reza Shah's modernization of agriculture through the encouragement of new crops like tea and sugar beet was effective and had a long-lasting impact.

The benefits of security were lost in the uncertainty of the war years. Some landlords and many bazaar merchants were able to make large fortunes out of the economic disruptions. For the most part, however, the countryside suffered political unrest in the tribal areas of the Zagros, in the north-west provinces of Azarbayjan and Kurdestan and in the remote villages where the rule of the central authority was weak. The difficulties faced by Iranian agriculture in the period 1900-46 were those of most developing agrarian economies struggling to modernize in variable political conditions in a generally harsh physical environment. But this position was slowly but demonstrably undergoing change. Increasing international interest in Iranian oil and a marked improvement in levels of oil income following the Anglo-Iranian oil crisis in the early 1950s were to introduce a new and growing impact of the oil sector on agriculture.

4 The Context for Change: Agricultural Development and the Move towards Land Reform 1946-63

4.1 Agriculture during the years of crisis 1946-54

Economic conditions continued to deteriorate in the period leading up to and during the 1951-3 Anglo-Iranian oil dispute. Development programmes were shelved since funding was not available. Any good intentions the National Front might have had, including land reform, were forgotten in the political chaos of Dr Mosaddeq's fight against the British and his struggle to maintain internal control against his increasingly strong domestic opponents.

Dedication to agricultural progress on the part of the authorities during the early 1950s is open to question. Admittedly, the confrontation with the Anglo-Iranian Oil Company and the British Government preoccupied those in power in Tehran. It has been argued that any incipient drift towards structural reforms in agriculture was pre-empted by the landlord class[1] which feared for its position as radical forces threatened to take over the state. Professor Lambton took the view that the governments of the Mosaddeq era had no ideologically radical or indeed practical dedication to improvements in the situation of the peasant cultivators.[2] Very few measures were put in hand to safeguard the peasants and those that were, such as the modification of share-cropping arrangements under the Full

Powers Act of October 1952, were heavily circumscribed in operation so as to leave the existing landowning conditions unimpaired.[3]

What was remarkable about the 1951-3 period was the almost complete neglect of agriculture by the authorities. Three factors are important evidence of this neglect. First, the country was embarking on an economic development plan that was almost entirely devoted to the improvement of water supplies and agricultural production when Dr Mosaddeq took over. Second, Mohammad Reza Shah had already inaugurated the sale of arable lands held as *khaleseh* on 27 January 1951 and had urged through the press that other landlords take similar action in breaking up their estates.[4] In effect, ministers in the Mosaddeq Government already had a precedent on which to implement a structural change in rural land ownership. Third, even without the plan, the very pressures generated by the cutting-off of oil revenues during the Anglo-Iranian crisis might have been cause enough for a crash programme for the development of agriculture, the one remaining productive, foreign exchange-earning and revenue-creating sector of the economy. That these considerable incentives were not successful in bringing the government to the conclusion that improvements in agriculture were necessary and urgent would seem to indicate that the regime was either blind to, or deliberately aligned against, policies that might bring about agricultural change.

The first seven-year development plan was adopted by the *majles* in 1949 and was, on the surface at least, directed towards development of agriculture, which was allocated more than 27 per cent of all public sector expenditures (see Chapter 7 for further details). During 1328 and 1329 (1949/50-1950/51) resources of the plan were absorbed in the establishment of institutional structures, including the Plan Organization itself, and the transfer to it of all state-owned industries. As the political upheavals accompanying the rising tide of events in the Anglo-Iranian crisis grew in intensity and as prospects of funds for the plan from the USA and from oil revenues diminished, so the developments provided for in the seven-year programme were allowed to lapse. The financial constraints on the government were enormous. In the event, barely 1,000 million rials was spent on agriculture during the seven-year plan, rather than the 7,300 million originally intended.

It is interesting, however, that it was not only foreign-exchange-funded capital works programmes that were abandoned during the 1951-53 period. Low rial cost projects were also suspended for the most part, and this is less easily justified. For example, the need for a much stronger information base was already established by the Hadary and Sai agricultural survey undertaken under the auspices of the US embassy in Tehran.[5] The pressing requirement for institutional reform in areas such as land ownership, and storage and marketing of farm products was also widely known. Iranian commentators listed these items among the main four obstacles to effective implementation of the second plan,[6] and they could have been mitigated or entirely avoided had the authorities persisted with the financially less burdensome components of the first plan during the 1951-3 period. Perhaps the Plan Organization's main contribution to rural development was the assistance given to the village councils created under the Full Powers Act of 1952, though the aid was not generous in its scope. While the early 1950s were difficult years for foreign agencies in Iran, it is notable that the United States Operations Mission, working under the US Point-IV aid programme,[7] was far more active in training and project implementation in agriculture than the organs of the Iranian Government. A principal cause of the government's failure to pursue the first economic development plan might well have been not so much the budgetary problems — though they were real enough — as the distaste of Dr Mosaddeq and many members of the *majles* for this essay in modernization that had been promoted so keenly by the shah. The plan's associations with the shah, with foreign interests and with borrowing abroad all served to reinforce those conservative attitudes which were apparently content to let the work of the Plan Organization run down.

Dr Mosaddeq and his colleagues had considerable problems in dealing with the distribution of crown land (*khaleseh*) provided for in the legislation passed in 1944 and 1946. The National Front's vague references to land reform before its coming to office should have satisfied its adherents that the small-scale and slow land distribution was worthy of support. In practice, there were genuine reservations concerning the status of those *khaleseh* lands, formerly private estates of Reza Shah, being sold off separately by the shah for the benefit of the Pahlavi Foundation, a charitable trust which managed the incomes from crown lands. They had been settled as religious endowments and were inalienable according to Islamic law.[8] Their sale was considered illegal and while it was doubtless made much of for their own ends by the large landowners, also gratuitously offended the religious establishment. It may be, as Professor Lambton suggested,[9] that the government was concerned lest the land distribution programme undertaken by the shah might bring him undue popularity. In the event, the sales were finally stopped in 1953 and the properties transferred into government ownership with the promise that the lands would be leased to the peasant cultivators. Following Dr Mosaddeq's overthrow, however, in August of that year, the lands were returned to the Pahlavi Foundation and the sales to peasants started once again.

There is no evidence that the government contemplated an augmented programme of reform of land tenure during the Mosaddeq years. With respect to water, however, an important piece of legislation was prepared under Dr Mosaddeq and ratified after his fall. This was the Law of August 1954, which extended the life of the Independent Irrigation Institute and laid down an extended set of rules for its operations which remained in force until the nationalization of water resources in October 1967.[10]

It became part of Iranian mythology that the country survived the oil crisis by returning to its agricultural roots. The reality was that shortages of all basic commodities were often acute, price inflation[11] was severe and unemployment was widespread. The political crisis denied the country the opportunity to reorganize its economy in a rational and gradualist way. Bharier summed up the economic disappointment of the period as follows: 'What has started as a "big push" to attain economic self-sufficiency thus ended as a feeble puff'.[12]

4.2 Monarch and *majles:* a struggle for power

Mosaddeq's fall from power cleared the way for the shah and the new government to return to planned development and recommence the albeit modest activity in crown land distribution. Indeed, a new second seven-year development plan was begun in September 1954 and the slow process of surveying the crown lands was continued. But the government was not single-minded in its concern with the economy. The country had to adapt to a new political balance and in particular to changed external circumstances in which the USA became the close ally of the regime and a powerful force for internal change.

Agriculture was a central political issue in the period up to 1963. The unsatisfactory status of the ceded properties, distributed by the Pahlavi Foundation, was used by the National Front to undermine the shah's legitimacy within a broader political field. Hints of new land reform legislation prompted many landlord interests to use the floor of the *majles* as a means of attacking him. A threat to *awqaf* (land endowed for religious purposes) in the early 1960s became the focus of religious opposition to the shah expressed through both the mosque and the *majles*. For their part, the shah and his allies exploited the land ownership question as a means of recruiting popular support against the landowners and the religious establishment.

It was perhaps inevitable that ownership of agricultural land was so important an area of political contention at this time. The result was not a concern for the farming community nor a real interest in agricultural development. All sides sought to use the agrarian question for their own political ends. Even the American agencies were involved in Iranian agriculture with transparent political motives. US representatives believed that the land reform would 'remove a source of popular complaint and help to counter internal communist propaganda'.[13]

The period 1954-63 was dominated by political conflict between the shah and the opposition on the issue of his management of government. The legislation that was so influential in creating the new structure of land holding and the institutional framework supporting agricultural activities was largely a by-product of this confrontation. Whether this situation was brought about by cynicism or the pressure of events, the result was the subordination of agricultural welfare to political expediency. Far from being a period of enlightened improvement of farming, the years leading up to the land reforms of the early 1960s were for the most part years of neglect of agricultural resources that continued rather than broke with previous approaches. The strong vested interests which had traditionally controlled social and political life in rural Iran through land ownership or trading activities were deeply entrenched. Unlike the opposition parties, they were, as a class, still widely represented in the *majles* and brought with them a resistance to change and an animosity to the Court which the shah's increasingly powerful security network could not entirely eliminate.

A number of notable victories were scored over the *majles* by the royal Court. The new international oil consortium was confirmed in possession of the southern Agreement Area, the zone generally corresponding to the former Anglo-Iranian Oil Company concession area. The National Iranian Oil Company was left with responsibility for nothing but 'non-basic' operations, which effectively removed the direct responsibility it had held from 1951 to 1953 for the running of all aspects of the domestic oil industry. Some reforms were initiated, but these were concerned with the break-up of the crown lands and required no legislation since they were in royal ownership.[14] A number of members of the *majles* contested the legitimacy of the sale of crown lands but they were ignored.[15]

The growth of US interests in Iran during this period was facilitated by the financial problems experienced by the British in the post-war period and by their preoccupation with adjustment to a prolonged process of decolonization in their overseas territories. The situation was exacerbated by the Suez crisis of 1956, which served both to diminish British prestige and to enhance the American role in the whole Middle East region. The principal agents determining US political supremacy in Iran were financial and material aid programmes and military assistance in the form of arms supplies and advisory services.

Within a short space of time during the 1950s, the United States established itself as an unrivalled influence at all levels of government with the deployment of its funds and personnel in support of the shah and his administration.[16] Military and agricultural agencies were the most visible and active US concerns in Iran at the time, both supported by aid programmes and a notable spread of regional bases throughout the provinces.

American financial aid was vital to the shah in the years following the settlement of the oil dispute. It took time for oil output to recover its pre-1951 level, while the government had adopted an ambitious national economic development plan, which required substantial aid from abroad.[17] At the same time, there was a vital need to rebuild the armed forces at some considerable expense.

American officials found the situation in Iran deplorable.[18] Reform of administrative, economic and social institutions was urged on the government as a means of improving efficiency, lessening the maldistribution of wealth and relieving class tensions. Progress was slow in achieving these aims. Not all Iranians shared the American view of the backwardness of their economy and society. Many were satisfied with the retention of attitudes and institutions which they knew and understood. Others, including landowners, clergy and bazaar merchants, had strong vested interests in retaining the *status quo*. For these and other similar groups increasing American involvement in Iranian affairs brought the threat of adverse economic change without any compensating political improvements. General disillusion with the American presence spread rapidly and almost universally throughout Iranian society as an expression of the disappointment of the high expectations entertained immediately after the end of the oil dispute and the apparent eclipse of British influence. A specific example of negative reactions to American-inspired reform is that cited by Okazaki,[19] who described the 1951 Decree for the Distribution of Crown Lands as implicitly an early American attempt to enforce land reform as a *quid pro quo* for continuing economic assistance. American pressures in favour of the distribution of crown lands resulted in the landowners and their allies in the *majles* uniting against the bill, and so modifying it in its passage through the assembly that it altogether failed to meet the requirements of either the government or the American groups which had urged the policy on the shah in the first place.[20]

It is an oversimplification to view the American impact on the modernization of the Iranian state and society as brash, inept and ineffective. Quite the contrary. During the five or six years during which the United States exerted a strong influence on Iran in the 1950s, initial steps were taken in what became a thoroughgoing, though especially rural, revolution in standards of lealth, education and transport, among other things. The great leap forward in hygiene and medicine that ultimately gave rise to the Iranian population explosion of the 1970s was made with considerable American assistance during the 1950s. Campaigns against malaria, smallpox and cholera were allied to the introduction of antibiotic drugs, whose main thrust originated in US aid programmes. Systematic improvement of other facilities, including the main and feeder road networks, also began at this time, funded and often organized by American interests.[21]

A less well-documented imprint of the American presence in the 1950s was left in the private industrial sector. The government made generous credits available through the banking system to those entrepreneurs willing to set up new industries. A significant number of light consumer goods manufacturing and assembly plants were established at this time, many using processes licensed by American companies or in association with US corporations (e.g. GEC, Jeep and others).[22]

The beginning of the 1960s saw the regime caught on the horns of a double dilemma. On the one hand, the economy was in considerable disarray. Foreign financial aid was promised but only on condition that a stringent economic stabilization programme was introduced that would bring the economy into balance, though at the cost of severe deflation.[23] On the other hand, political relations between the monarchy and the national assembly had reached an impasse that could be resolved either by the shah acquiescing in rule within a constitutional monarchy or by the effective elimination of the assembly from the processes of government.

The shah chose to solve both his problems by the appointment of Dr Ali Amini, who was to act as prime minister while the assembly remained suspended, and was charged with responsibility for righting the economy and ending political unrest within the country. Dr Amini had been trained in the United States and was widely believed to have US support. He had the reputation of being independent and far from overwhelmed by the authority of the shah. As a major landowner from the north of the country, he had considerable respect. His administrative talents were admired outside Iran and there was confidence, particularly in the USA, that he would be able to restore good management to the national budget without precipitating a political crisis.

Amini ruled by decree and set up a council of Ministers consisting of a strong group of individuals, many dedicated to reform, such as Hasan Arsanjani, the Minister of Agriculture. This administration was generally believed by Iranians to be endorsed by the United States in so far as Dr Amini was seen to be its nominee.[24] The performance of Amini and his colleagues was extremely efficient. An economic recovery programme was begun and the country's financial affairs were put in order as demanded by the International Monetary Fund, from which Iran hoped for further loans to tide it through the second development plan.[25] Among the several measures taken by the Council of Ministers was a decree calling for land reform on the basis of the 1960 legislation originally passed by the *majles*. This became the basis for the far-reaching agrarian reform during the years 1962-68 and in many ways prepared the way for the shah's 'white revolution' programme of 1963.

Politically, however, Dr Amini and his cabinet were too complex and potent a mixture for the shah to coexist with. They offered the temporary compensation that government could continue without the national assembly, but they were independent-minded, had plans for modernization that were far from the shah's liking, especially where they touched upon constitutional arrangements, and had direct links with the United States which might weaken his personal position. It was clear, too, that elections would need to be held if the terms of the constitution were not to be strained beyond breaking point. The shah seems to have been altogether unwilling to contemplate Dr Amini contending for power in the elections from the position of incumbent premier. Amini and the Council of Ministers resigned in 1962 and Asadollah Alam was appointed by the shah as the new prime minister.

The brief premiership of Dr Amini excited somewhat polarized responses. He could never be forgiven by the National Front for agreeing to govern without the national assembly. He was vilified as a mere appointee of the shah by the religious establishment. He was deeply suspected by nationalists and the groups of the political left as a place-man of the Americans. In retrospect, the Amini administration might be regarded as another missed opportunity for the creation of a workable form of representative government, sacrificed by the shah on grounds of dynastic expediency. Only the land reform, begun by Dr Arsanjani as Minister of Agriculture, remained as a monument to the often imaginative and serious work started in this period, though other proposals for change were later adopted by the shah in his 'white revolution' programme. Certainly, the Amini Government acted as a stop-gap between the unhappy attempt at constitutional monarchy in the second half of the 1950s and the personal rule of the shah in subsequent years.

Elections were held in September 1963, and once again denounced by the opposition as government-controlled and undemocratic, which they most probably were. The *majles* did, none the less, contain a new mixture of deputies, with a large proportion of young technocrats, many of them associated with the Land Reform Organization or similar movements. Officially, the Mardom Party formed the government with Asadollah Alam as Prime Minister; his party had no overall majority in the lower house but survived with the support of independents and lack of challenge from elsewhere.

A wide range of basic reforms, many of which were prepared originally under the Amini administration, were adopted by the shah's advisers as a set of guiding principles within a six-point reform programme named the 'white revolution' (*en-*

qelab-e sefid) or the revolution of the shah and people (*en-qelab-e shah va mellat*). Originally consisting of land reform, a literacy campaign, electoral reform including the enfranchisement of women, nationalization of forests, profit-sharing for industrial workers, and the sale of state-owned factories, the programme was clever in conception and was designed, or came by degrees, to fulfil a number of functions. Indeed, the political leverage offered by the programme was possibly ultimately more important than the actual content of the individual reforms.[26]

The reforms were widely recognized both at home and abroad as vital and desirable. They benefited, or had the potential to benefit, large numbers of people ranging from farmers to urban workers. Efficiency in the management of forest resources and the old industrial plant was to improve as the forests were taken from private owners and the industrial plant moved from the care of the bureaucracy to private hands. There might have been some cynicism, too, in the fact that most of the measures included in the programme incorporated the reforms proposed by the National Front, leaving it in the awkward position of having to attack policies that were formerly integral components of its own platform. An important facet of the six-point reform was its value in showing the regime in a favourable light to liberals in the United States and Western Europe, where the shah was increasingly portrayed as an enlightened modernizer of a backward country struggling against reactionary forces.[27]

Given so unimpeachable a programme of change, which had the advantage of expresing the shah's concern for the improvement of his country, the regime was able to recast the relationship between assembly and monarch. In explicit terms, the assembly became the instrument for effecting the shah's reforms and its innovative or opposition role was curtailed, if not abandoned. Prime Minister Alam declared himself as the servant of the shah even before the 1963 elections, and publicly stated that he saw his task as the implementation of the six-point reform.

Promulgation of the reform programme and the political sleight of hand that was designed to give the shah all but complete control of the country did not pass without a variety of challenges from other power groups and vested interests. Fortunately for the government, those opposed to the 'white revolution' were divided, and the authorities were able to pick them off one by one. Landlords who were adversely affected by the agrarian reform legislation of 1962 (see Chapter 6) or who feared for their future under an extended government operation bent on destroying their economic power and influence in the countryside, attempted to form a 'Landlords Association'[28] to dissuade the government from further action. The authorities bussed farmers and others into Tehran for a so-called farmers' day, a meeting which vociferously rallied to the defence of the land reform and served to warn off the landowners, many of whom were in any case subject to personal intimidation to persuade them to withdraw their public opposition to the reform. Among the landowners whose teeth were not drawn at this time were the clerics and guardians of shrine properties.[29] They had both ecclesiastical and financial reasons for disliking the reform which they did not hide,[30] even though the first phase of reform did not affect *vaqf* land. At the same time, many of the clergy were landowners in their own right and were antagonistic to the reform. Many theologians, however, had valid religious grounds for questioning the legal and religious validity of a law that seized private property, and these feelings were equally strongly mobilized against the land reforms attempted after the 1979 revolution. The *'ulama* chose, however, to subsume these objections within a broader confrontation with the shah's government which did not break out until after the landlords' revolt was defused.

In the Zagros Mountains region the regime faced a growing tribal revolt among the Qashqa'i people. Resentments against the central government had simmered in the area for many years, breaking out into open rebellion during the Second World War.[31] The return to the tribal territories of the traditional leaders of the Qashqa'i, including Naser Khan, gave strong leadership, which mobilized a variety of chronic resentments on the part of the tribal peoples to spill over into a full-scale revolt. Gendarmerie units were driven from the area and the writ of the central government was all but ended in the highland zone south of Shiraz.

The most serious unrest came in June 1963. On 5 June there were demonstrations against the regime on the streets of Qom and Tehran, with other centres, notably Mashhad, also becoming involved. Opinion is deeply divided on the motivations for and the course of events. It has been suggested that incitement to revolt came from the religious hierarchy which was persuaded that the actions of the shah in dissolving the *majles* and ruling by decree were tantamount to usurpation of proper authority and within the domain of oppression.[32] Undertaking measures that were deemed unislamic by the Shi'i hierarchy aggravated this situation of conflict between the shah and the religious leadership.

A number of specific complaints were made against the regime. The two that most appeared to catch the popular imagination were the questions of land reform and the enfranchisement of women, though a general antipathy to the regime and its foreign allies was noticeable. Rioting was also used as a cover for activities against the regime by some landlords and merchant groups in the Tehran bazaar.

The government took a firm stand against the unrest. Troops were brought in to suppress the demonstrations[33] but with a high death toll, variously reported as 'several hundreds' by the authorities or 'many thousands' by opposition sources.[34] Prime Minister Alam calculated accurately that the government could not afford to let the rioting get out of hand or run on, thereby permitting other anti-regime groups to take advantage of the deteriorating situation. His prompt destruction of opposition on the streets gave the shah a notable victory, the effects of which were to last for some fifteen years. Following the defeat of the religious revolt, the shah was virtually free to rule the country on his own terms. Even the causes of the demonstrations — objections to land and electoral reforms — were turned to advantage in order to exhibit the regime as a modernizing administration overcoming medieval obscurantists seeking to block its path.

There were relatively high costs for Iran in the events of 5 and 6 June 1963. The shah was confirmed in his view that he could afford to ride roughshod over the opposition, whatever its source. From mid-1963 there were few occasions when he appeared to heed advice on questions of political liberalization. Indeed, his belief that the organs of state should be designed to serve his requirements was increasingly underpinned by the extreme flattery of politicians whose only source of authority derived from His Imperial Majesty. Yet much of the success in overcoming the religiously inspired riots belonged to Asadollah Alam, and the basic political programme, including land reform, which sustained the shah was taken from the Amini administration of 1961/62.

The implications for the country of the events of June 1963 were to be considerable. It has become the conventional wisdom that June 1963 marked the entry of Ayatollah Khomeini into a leading role within the religious hierarchy, which in turn cast the die for the flowering of religious opposition to the monarchy that culminated in the revolution of 1979. Such a view must be regarded as partial. While it might be accepted that Ayatollah Khomeini fully emerged in 1963 from the shadow of Grand Ayatollah Borujerdi and did indeed play an important role in the organization of the June riots, his disappearance into exile in 1964 brought a reduction in his ability to influence Iranian political life. He none the less maintained his position in the minds of many Iranians through the distribution of written and recorded versions of his teachings and utterances, which were an important prop for his dedicated supporters in the religious hierarchy inside the country.[35] What must be seen as far more important in the context of the actual course of events in the medium term inside Iran were the emergence of the shah as undisputed ruler and his freedom to adopt policies for economic change and development. The imposition of fifteen years of autocracy from 1963 to 1978, within which there were five years of precipitate economic growth in the years 1973-78, might be perceived as instrumental in creating the conditions for revolutionary upheaval and for the return of Ayatollah Khomeni as much as anything surrounding the Ayatollah himself during that period.

Abrahamian's view that the 'Muharram upheavals of June 1963 were to be a dress rehearsal for the Islamic revolution of 1977-9'[36] contains obvious elements of truth, but it was equally important for the future of Iran that the shah set in train a radical modernization of the country that ran — often with great material advances — until the late 1970s. It remains too early to judge whether the social and economic changes of 1963-78 will ultimately leave more or less impression on the character of the nation than the political revolution of 1978-9.

4.3 Concentration of political control under the shah: the context for the agrarian reform

By 1963 the reforms of the sixth Bahman 1342 (26 January 1963) had become the principal elements in the government programme. Agriculture was the main continuing component and it might have been supposed, on the basis of the regime's apparent dedication to agrarian reform as demonstrated in the firm handling of the 1963 riots, that it would have strongly supported the objectives of the original land reform. In reality, however, agriculture lost considerable political ground after 1963. Amini and Arsanjani had both left office and with the loss of the latter in particular the enthusiasm and crusading spirit of the land reform were dissipated. The new men who came in with the elections of 1963 had no political capital invested in the reforms. They took office merely to implement the provisions of existing laws within the policies laid down by the shah and with some, if not much, respect for the views of the young deputies who had come into the *majles* from the Land Reform Organization.

Agriculture was useful to the government, though increasingly as a tool for winning political popularity with the peasantry and those urban residents not committed to the left or the National Front.[37] Indeed, land reform was still seen as a means of undermining the landlord class and obliquely as a lever against the National Front,[38] some of whom were from, or had close relations with, the former land-owning class. Above all, land reform was a powerful argument abroad in demonstrating the regime's concern for the mass of the people and for improving the standards of living of the peasants.[39]

Behind what was a thin pretence of commitment to the original principles of land reform, other factors were at work.

The process of centralization of decision-making and activity in Tehran was marked.[40] Rural communities appeared to have a decreasing say in the events affecting them. Organs of the central authority intruded into all the villages as improved surface and telecommunications links made possible a tighter control of the countryside.[41] As part of the same process there was a growing concentration of authority in the hands of the shah and an increasingly authoritarian aspect to the state in its contacts with farmers and farming. Notions of the flowering of an independent rural peasant farmer class as a political force within the country, which had been envisaged by Dr Arsanjani, completely faded.[42] Even the organization of democratic rural co-operative societies and unions was dropped in favour of direct orders from the centre.[43]

Agricultural policies after 1963, as will be seen in Chapter 6, were conceived, planned and administered from the capital. There was a widespread feeling in the villages that the government was out of touch with the sentiments of the rural population and that the promises of the early reform period were being reneged on. These attitudes were reinforced by the watering-down of phase two of the reform,[44] by the shortages of farm credits and by bureaucratic interventions in village life. The disenchantment of the peasantry with the government was not immediate but was a perceptibly growing trend in the years 1963-67, after which alienation was clear and an important factor in retarding investment in the traditional agricultural sector.

The course of political change during and after 1963 was to have an especially strong impact on agriculture. Many of the changes arose from the failure of the opposition to land reform and the Sixth Bahman programme. The position of the religious leaders was weakened following the exile of Ayatollah Khomeini, and the old middle class declined in influence as the landlords lost their control of the *majles*.[45] New political interests emerged in the shadow of the shah's Court, with policies and economic priorities very different from those of their predecessors. The period 1963-67 saw a gradual clustering of technocrats within a stable and self-confident administration whose approach to agriculture was governed by target-achievement initially expressed in implementation of the existing land reform but later manifested in new aims reflecting needs in areas of the economy other than agriculture.[46]

The Mansur administration, 1963-65, was unfortunate in taking office at a time of severe economic depression. Three to four years of economic decline had brought a measure of privation to many. Unemployment was widespread in the urban areas, while several rural districts, notably in the east of the country, were adversely affected by persistent drought conditions.[47] Efforts to reform the administration of the country together with moves to reinvigorate the economy were just beginning to bring better economic prospects by the second half of 1964. In many ways, 1965 marked the beginning of a remarkable period of prosperity, though in the early months of that year this was not visible to most Iranians.[48]

Amir Abbas Hovayda, Hasan Ali Mansur's successor as prime minister, adopted a passive attitude towards the constitution and the *majles*. He aided the emasculation of the assembly and assisted the channelling of power into the hands of the shah. The reform programme, which had been significantly elaborated by the creation of new agencies and imaginative initiatives to cure some of the country's most backward features, such as the Literacy Corps as a means of reducing the poverty of school facilities in the rural areas, became blunted with the passage of time. Additions to the programme were made but lacked the enthusiasm, relevance and conviction that had characterized its adoption earlier on.

The concentration of power in Tehran and the coming of age of the young technocrats during the second half of the 1960s indicated an adverse trend for agriculture on a number of grounds. Those in power were unsympathetic to the problems and anxieties of the farming population. Very few members of the cabinet were from farming backgrounds and the technocrats who supported them were often better travelled in Europe and the United States than in Iran.[49] Those in Tehran specifically responsible for the land reform, agriculture and water supply were army officers or engineers. In the Ministry of Economy a dynamic management team was dedicated to setting up a new economic structure that would enable Iran to become as diversified as possible away from oil dependence.[50] Their interests lay in rapid industrialization and economic modernization in which agriculture's role was important but subservient to the needs of other sectors. Rural areas were seen as the supply point for foodstuffs, basic exports and cheap labour and as the sources of demand for Iranian industrial and other products. The problem was not the inadequacy of what were imaginative and useful economic programmes but the assumption that agriculture and the rural sector could take the strains imposed upon them by the new policies. It was taken for granted that agriculture was so established and monolithic a system that it could weather rapid change and a degree of over-exploitation, while resources were channelled to the development of infant industries and services in the towns and cities. Events were to show that Iranian agriculture was far from indestructible and that modernization of the urban sector was achieved at the cost of neglect of the real needs of the farming community.

5 Water Resources and Irrigation Cultures

5.1 Introduction

By Middle Eastern standards Iran is a country well provided with land and water resources, yet only a tenth of the surface area is cultivated in any one year. The areas of arable farming correlate closely with the availability of water supplies, whether provided by natural precipitation or underground reservoirs. This chapter will examine the strengths and weaknesses of the country's water resources in order to substantiate the argument set out in Chapter 2 that the environment is for the most part hostile to agriculture, which has survived by ingenious adaptations to very varied geographical circumstances.

The limited reclamation of land to arable use is remarkable in view of the general assumption in Iran that the national endowment in soil, water and climate is rich. Estimates of land use are poor and for the most part unreliable. A serious and useful first approximation of the areas under agricultural use was made by Hadary and Sai[1] under the auspices of the US embassy in Tehran in 1949 as a basis for the formulation of the economic development plan by Overseas Consultants, Inc. Most later data on land use for the Second and Third Plans were calculated on the basis of the Hadary and Sai study.

Despite the fact that the figures are offered as no more than rough estimates, the core material was seen as reliable, including the area for grains and other field crops. The intensively cropped arable area in 1948/9 is encompassed by these two categories to give a total area of 3,920,000 hectares or less than

3 per cent of the country under the plough in that year. At that time arable farming provided slightly more than 40 per cent of agricultural incomes across the rural population, with livestock providing approximately the same proportion. The balance was made up by timber, fruit and nuts.[2] The role of a relatively small area of land in providing the mainstay of the country's agricultural output was reinforced by the finding that one-third of the grain and two-thirds of all other field crops came from the irrigated farms which occupied less than 2 million hectares. Much of Iran's struggle in the period since 1949 has been in improving the amount and rate of utilization of its arable land.

Wresting gains in this arena proved difficult. Taking the 1949 study by Hadary and Sai as a base year, and adopting extrapolations calculated on additions to the cultivated area arising through projects reported as complete by the Plan Organization, the Ministry of Water and Power and other agencies, an attempt has been made to plot the changes in agricultural land use for the period 1948/9-1972/3. There were notable improvements claimed, especially in the area of irrigated land, which all but doubled to some 3 million hectares as both new land was brought under irrigated cultivation and former drylands were converted to irrigated culture. Some qualitative improvements were also made as semi-irrigation was replaced in some regions by full irrigation.

But these trends were less impressive when measured against rising population numbers. The availability of cultivated land per head of total population fell consistently in the twenty-four years ending in March 1973. There was roughly a hectare of agricultural land defined in the broadest way per person in 1948/9 but only two-thirds of a hectare by 1972/3. The ratio of irrigated land showed a similar trend. If the rural population alone was taken in relation to the availability of land, the situation deteriorated perceptibly for total agricultural land, but less so for irrigated land.

A series of new estimates of the land-use pattern was undertaken in 1974/5 by the National Cropping Plan[3] and by the agricultural census of 1977/8. The findings of the two were not in accord on all points but the essential components for the total cultivated and irrigated areas were close enough to confirm a new base level for land use (see Figure 5.1), though one not comparable with earlier figures because changes were made in definitions and techniques of assessment between the two periods. The National Cropping Plan in particular was based on a survey by remote sensing and other technology not available in 1948/9 and had the advantage of greater accuracy over previous estimates. Many scattered regions of cultivation which were badly surveyed under the earlier studies were located and measured in the 1974/5 evaluation, though this itself was also subject to inherent distortions, and cannot be regarded as a final and definitive statement on the land-use pattern of its time.

The land/man ratio improved as a result of the recalculation

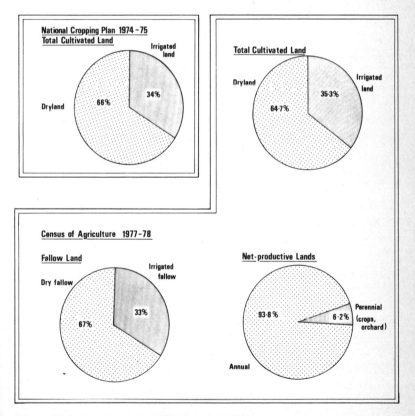

Figure 5.1. Land use in 1974/5 and 1977/8.

of land use in the mid-1970s, though only in the irrigated sector, where it was twice the level recorded in 1948/9. Taking the population as a whole and the area of cultivated land, the ratio continued to fall. Relative intensity of use of farmland rose against the livestock sector and by 1977/8 arable farming provided about two-thirds of agricultural income and livestock one-third. Declining output of livestock from the tribal areas doubtless made some contribution to this change, while the move away from subsistence agriculture and a measure of farm mechanization added to pressures against the role of livestock in rural areas.

The gradual improvements made in use of available water supplies were important in so far as they enabled an expansion in farm output from increasingly reliable sources. But the efficiency with which the potential water resource was converted to effective agricultural use was woefully behind expectations. At the beginning of the 1950s it was calculated[4] that annual consumption of irrigation water was 28,000 million cubic metres. This total rose slowly during the Second and Third Plans to stand at some 29,000 million cubic metres at the end of the third plan period. As a result of the concentration on dam construction in the 1960s new water resources became available, lifting the total amount of irrigation water used to some 33,000 million cubic metres by 1973. As we shall see later, the administrators of the Fifth Plan envisaged a large expansion in water storage and provision for irrigation to more than 50,000 million cubic metres. This was not achieved, though some progress was made as programmes begun in the Fourth Plan came to fruition, possibly taking in a gross volume of 10,920 million cubic metres.

The volume of water available for irrigation was actually far less than the gross figures above suggest. Urban water use rose rapidly throughout the period 1950-78, particularly in the last decade as the size of the urban population grew and as demand per head also increased. There were additional claims from industry for more water. The water resource also came under pressure from the surge in the electricity requirement in the same period. Approximately 4-5 per cent of all Iranian electricity supply came from hydroelectric power stations, which were important contributors to the national grid during peak consumption. Although there were some reregulating works below the main dams, much of the water run through the turbines at the hydroelectric stations in the cold season was then lost to agricultural use in the summer irrigation season.

5.2 The development of water laws

There is no evidence to suggest that water laws retarded the development of water use in agriculture, though water law in Iran is complex and has important regional variants. Islamic Law and the Civil Code were jointly the major influences on water use until 1979. The Shi'ite interpretations of Islamic Law as it affected man-made water provisions by *qanats* and artificial canals were fairly direct and unequivocal, in so far as the owners of water sources had absolute rights over water use vis à vis other individuals and the community as a whole.[5] In the case of rivers, streams and other natural water courses, ownership was vested in the community but in practice upstream riparian landowners took precedence in the use of water over those downstream, and established users took priority over newcomers. But there were provisions in the law that offtake should be limited to prescribed limits by each landowner in turn.[6]

The investing of the individual with rights in water ownership and usufruct, whether from man-made or natural sources, meant that water became subject to the laws of inheritance. Fragmentation of irrigation water supplies in villages that were largely made up of small landowners and cultivating peasants became severe. In the cases of *qanats* that suffered from markedly reduced flow in the dry season, as was a frequent occurrence in many villages visited by the author in eastern Iran, the summer supply per water share from *qanats* with highly fragmented ownership enabled very few farmers to cultivate commercial dry season crops. Most cultivation in villages or parts of villages of *qanat* service areas was restricted to kitchen-garden cropping in the central cluster of walled gardens. Co-ownership of water supplies gave rise to some difficulties over care and maintenance. Criticism of *qanat* management and failure to motivate joint action to repair channels that were deteriorating in output were not uncommon in instances of this kind.

The Civil Code to a large extent encapsulated many of the ideas and rulings of Islamic Law. As Ghahraman has pointed out,[7] the Civil Code had confusing and contradictory provisions on the matter of man-made water sources. It was, however, much more precise in respect of natural water channels and, following Islamic Law, gave priority to established owners of land over newcomers and upstream over downstream users of water. The principle that the irrigation interests of others shall not be infringed by newcomers was enshrined in the legislation within the Civil Code, which also confirmed that priority to water use was determined by precedence in time rather than by proximity to the water source. Thus, an established farm at a distance but drawing water from a river or stream would have prior access to its usual volume and timing of water supply over a newcomer situated on the bank of a stream.

The very nature of water provision and irrigation systems in Iran, many of which were fragile and subject to rapid deterioration, meant that the corpus of laws was arranged to accommodate their special needs. The *harim* or prohibited area for water extraction around an existing well, spring, *qanat* or canal was designated as reserved so that their productivity should not be impaired. This was achieved by forbidding any works that might damage the water source. Specific distances were laid down, including, for example, 30 *gaz* (approximately 30 metres or 100 feet) within which no third-party interference was permitted for an irrigation water well, and the general principle of uninfringed existing rights prevailed under the Civil Code. Implementation of the law was a different matter. It was not unusual for powerful individuals to sink wells and, using powered pumps, to trespass on the *harim* of other well systems, especially those serving *qanats*. Cones of depression within the aquifer created by the new water-lifting systems effectively deprived *qanats* and shallow wells of their water source. The example of Bayazeh in the eastern Dasht-e Kavir, visited by the author in 1970, was one where competition between water-lifting devices that ignored the laws on *harim* led to a general fall in the local water table that ultimately damaged everyone's interests. A similar situation existed in other areas, especially those located around the Dasht-e Kavir.

Provisions in the Civil Code for the care and maintenance of water supplies had obvious importance in semi-arid and arid areas where sand fill and silting were common in canals and *qanats*. Roof falls and silting in *qanats* were also frequent, occasionally exacerbated by flash flooding and earthquakes. It was laid down that all disputes over repairs to water sources were to be referred for arbitration to the courts under clauses that enabled financial sanctions against recalcitrant co-owners who failed to participate in refurbishing the system. All too often the legal authority was inadequate to enable remedies to be made rapidly and effectively. The co-owners were often unable to overcome their inertia or an influencial co-owner blocked the moves to bring the matter to judgement. In this way many *qanats* and irrigation channels became blocked and inefficient.

Although Islamic Law played a key role in water rights under the monarchy, and in the contemporary period theoretically governs all regulations in this area, the reality was that day-to-day running of the irrigation system lay in the hands of village officials occupying traditional roles and interpreting management of water on the basis of rules and measurements that were archaic. General legal frameworks from Islamic Law and the Civil Code that surrounded water use were powerfully supplemented by customary practices (*'urf*). The origin of many customs lay in pre-Islamic times and was local rather than central in administration and practice.[8] These local regulations governed to a large degree the access to and use of water in irrigation within what was a complex organization of supply in an uncertain physical environment.

The survival of customary practice through into Islamic times appears to have been considerable according to Lambton,[9] not least since Islamic civilization was urban in orientation and, by implication, was less concerned with the minutiae of activities in agriculture or pastoralism. The special position of customary laws was recognized quite specifically in the Civil Code adopted by Reza Shah in the modern period. In effect, customary law was left to apply in all situations for which special legal provision had not been made elsewhere. Given the widespread survival of the *qanat* and other traditional water sources, *'urf* remained the residual body of regulations for water use in most areas and for most types of water supply system until the present time.[10]

In addition to the gamut of regulatory principles governing water rights, governments sought to set up an 'improved' and more relevant code for irrigation. Much of this was lost as a result of political problems. The legislation under Reza Shah included the *Qanat* Law of 1930 for the protection and construction of *qanats*, and the law of 1937 for the development of village land and water resources under the auspices of regional councils. Neither was pursued with vigour, though the 1937 law, given adequate funding and political stability, might ultimately have had a beneficial effect. As it was, the inefficiency of the civil service and the ability of the major landowners to monopolize the financial resources put at the disposal of the councils made the measure largely ineffective. Formation of the Independent Irrigation Institute in 1943 was a far more significant step. It provided a national organization to prepare background information and proposals for the conservation and development of water resources. At the same time, the statutes of the Institute were seen to violate some of the principles of Islamic Law since they overrode private rights in water. In 1954 the Institute began the task of assembling data on stream flow, water ownership and irrigation systems that were to be vital in the later development of storage dams and other construction works on the country's rivers.

The land reform and the nationalization of water resources were far-reaching in their effects on irrigation. In the land reform, albeit at a late stage, water ownership was transferred from one group to another as the land it served was distributed. Not officially part of the reform but an important abutment to it was the development of irrigation works in the service area of the major dams by the Ministry of Water and Power under the agribusiness programme. Nationalization of water was introduced in 1967 as the tenth point in the 'white revolution', and confirmed by royal assent in July 1968. It provided that the

Ministry of Water and Power should protect and manage the use of all water resources. Regional boards were established by the ministry to assess the volume of water use, to take powers to restrict water consumption and to charge water rates.

The 1968 legislation took a great deal further the powers that had originally been granted to the Independent Irrigation Institute. The ministry was vested with very wide powers of control, administration and taxation for water of all kinds, that clearly devalued the status of private ownership. Many ideals of Islamic Law were also overturned or ignored. There was little doubt, however, that the legislation was needed. The application of a monetary charge to water use in areas where dams, rivers and stream sources were exploited gave an incentive for economy in its consumption where wasteful application had formerly been thought to be the case.[11] Allocation of water between different crops, particularly between commercial and subsistence crops, would have become much more finely tuned as water charges took effect, with wheat and other low-value crops giving way to cash crops.

It was also argued that the state could not be expected to invest in water supplies from major dams, provide funds for new local water systems, and give credit to improve traditional irrigation works without some reasonable return in the form of current income from water charges. In the period before the nationalization of water the financial incentive for the state to invest in irrigation facilities was low. Not only did agriculture imperceptibly increase its value-added, but no direct income accrued from the sector to official organizations. In the case of existing water storage structures, the managing authorities found themselves far more inclined to use the water for purposes that would produce income through block sales of electricity by letting water run through the turbines at the dam site. The supply of water to urban areas similarly offered a more certain financial return for the Ministry of Water and Power than release of water to agriculture. After nationalization some of the disincentives to the official agencies of supplying water to the irrigation sector were removed, and to this extent it was a beneficial step.

Misguided thinking was also at the heart of the water

nationalization law. It was a more or less consistent article of faith that agriculture in traditional village areas was primitive and that irrigation practices were antiquated and inefficient. Even the water systems, especially the *qanats* themselves, were disparaged,[12] and efforts were made by the planners for more modern water-lifting techniques to be employed.[13] In fact, there was every reason for caution in the spread of well drilling, given the poor state of knowledge of the sub-surface aquifers, not least since pumping tended to render *qanats* in the same area subject to rapid obsolescence, while wells in close proximity could push down the water table extremely, as at Bayazeh and other settlements adjacent to the central deserts.

Assumptions that irrigation practices were less than economical in their consumption of water were not founded in any understanding of real local needs. Prescriptions for the amount of water to be used, most of which were legal in origin, coincided with long-standing agronomic standards[14] recorded in the literature. More importantly, there were often very good environmental reasons why heavy irrigation was widely practised.[15] The shares in a water supply expressed the water rights of each landowner and did not necessarily conform to the water requirement of the farm in question. In circumstances where the rotation might not return the water supply until after the optimum time for the crop in question, farmers would soak their land in order to give the crop the best chance of survival. The water rotation was on a night-and-day basis so that irrigation had to take place during the periods of greatest heat — a situation that called for heavy irrigations to prevent shallow watering and root damage.

For much of the Iranian plateau and the southern plains heavy irrigation had the advantage of carrying down natural salts that might otherwise have concentrated at the surface and induced high levels of alkalinity or salinity. In the east of the country particularly, though it was a more general problem from time to time, drying out through strong and persistent winds was a major problem that could be offset in part by ensuring thorough watering of crops. Rapid penetration of hot dry and often sand-laden air in the south-west which reduced relative humidity down to low levels presented the same problem to farmers and was answered by heavy application of water during irrigation. It was a notable feature of traditional agriculture in Iran that few areas were permanently lost to agriculture through soil salinity, which in itself was a tribute to established irrigation practices. Problems of salinity were soon experienced in Khuzestan in the 1970s in the agro-industrial areas where modern irrigation systems were used, though those very same lands had formerly been cultivated for many generations by local and itinerant traditional farmers.

5.3 The irrigation regions

Irrigation was the mainstay of agriculture but for much of the country the available water resource was stretched between forms of full perennial irrigation, seasonal irrigation and semi-irrigation. There was great variation from region to region and village to village in the combination of types of irrigation and dryland cultivation. Two critical environmental influences were the rainfall regime and the accessibility of other water supplies. A brief introduction to the physical geography was given in Chapter 2. In this chapter the position will be elaborated to establish a series of irrigation regions. Criteria for isolating each region will be irrigation type as defined by water source, intensity of irrigation and the comparative role of the dryland sector.

Figure 5.2 attempts to identify the principal drainage/irrigation areas. Most characteristics of traditional village irrigation cultures were associated with more or less definable regional bases and whereas there were variations within each region and a generous incidence of eccentric villages that deviated from the standard pattern, they are insufficient to undermine the overall picture. Many, though not all, modern irrigation systems show less regard for the underlying traditional structures and determining factors of local environments. In the case of the Sefid Rud dam, the irrigation waters ran finally into the established irrigation and field systems of the Caspian Plain, thus causing few difficulties of classification. On the other hand, modern well field water supplies in parts of Fars offered an entirely different commercial and topographic presence from the tradi-

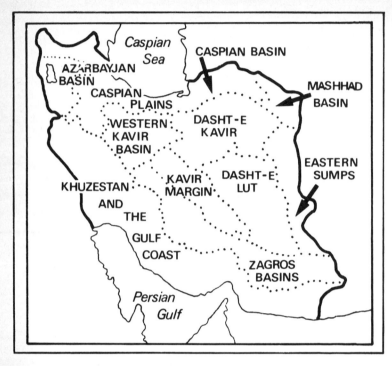

Figure 5.2. Irrigation regions.

tional villages of the region. It might be hazarded that one cause of the high failure rate among the irrigation/farming schemes introduced in the modern period was their inability to take advantage of traditional local skills or adapt to the demands of the local environments. The distribution of the various irrigation cultures of the country is important to an understanding of the regional nature of agriculture as the following examination will show.

5.3.1 *The peripheral basins and plains of the central plateau (Western Kavir basin and the Kavir margins)*

The dominating irrigation culture was that of the *qanat*. Its use was widespread across the Iranian plateau, where it provided water for perennial irrigation in the great topographic and internal drainage basins that lay between the central deserts and the mountain chains surrounding them. Here many *qanats* were constructed with the object of giving (and they still provide it in many cases) strong water flow throughout the year. On the mountain peripheries the water tables within each basin were kept reliably replenished by rainwater but above all by snow melt seepage so that the wells feeding *qanats* could be driven into the permanent water table, thereby ensuring in the most favoured cases an undiminished supply regardless of the season and variations in rainfall. This so-called *gharq-ab qanat* or *ab-ghari* was found most frequently in the great basins that lay around the central plateau, as a result of the nature of the water seepage and the structure of the quaternary strata in them which permitted the accumulation of underground water in large, often deep and undisturbed, sedimentary layers trapped between impervious beds.

For the most part, however, *qanats* were subject to seasonal and year-to-year variations in their output in response to the replenishment or depletion of the aquifer. Such *qanats* were known as *hava bin*.[16] In areas of the Dasht-e Kavir *qanats* with a close response to current rainfall conditions were designated as *pusht-ab*. Water supplies, principally of the *hava bin* kind, though ranging in reliability of production through to the *qarq-ab qanats*, were the basis for cultivation in, for example, the agriculturally rich areas around Esfahan, Yazd, and Kerman. Water supplies were supplemented where possible with river water as at Esfahan, but the *qanat* was the mainstay of the system.

5.3.2 *The central desert areas (Dasht-e Kavir and Dasht-e Lut)*

The *qanat* also played a significant part in water supplies elsewhere. In the central deserts and the smaller basins within or surrounding them it was traditionally the single source of irrigation water. Water tables in these areas were notoriously less reliable in their yield than those in the higher basins closer to the mountain chains that lay on the perimeter of the plateau. Not only were they less prolific but they frequently depended on very variable and scanty rainfall within a limited catchment area for their replenishment. Because most of the sub-surface aquifers in or near the deserts suffered from problems of high evaporation and low recharge, they were affected by salinity so that gathering systems for the *qanats* were impeded by shallow access to the water table in a number of areas, often limited to the upper phreatic aquifer, which aggravated problems of reliance on replenishment of the reservoir from rainfall.

Qanats which drew from the upper water table were known as *pusht-ab* and their output ebbed and flowed in relation to the annual rainfall.[17] The marginal nature of the *qanat* systems in the central plateau region was demonstrated by the constant change in the siting of villages and the cultivated areas they served. New villages arose such as Chupanan north of Na'in, where the author was told that the first major expansion there came at the end of the nineteenth century. By 1963 there were three flourishing *qanats* at work providing what was described as plentiful water for barley, wheat, fruit and date palms. Not far away in the Khur district, there were the remains of abandoned villages such as Ghaffurabad, where the *qanat* had gradually dried up. Mahdavi found a similar rise and fall in cultivation in the Abuzaydabad district in the western Dasht-e Kavir.[18]

The effects of a water supply coming mainly from *qanats* which fluctuated in output between years and seasons inhibited agriculture considerably. Summer cropping was confined to small gardens where fruit trees and a modest area of vegetables could be cultivated. Expansion of the water supply was expensive and difficult before the coming of the diesel pump and drilling facilities for deep wells so that the emigration of young men away from the desert villages was a permanent feature of life there. At Jandaq, where the two *qanats* serving the main cultivated area were low yielding during my visit in 1964, most young males had moved to Tehran, one family, for example, concentrating on laundry work in the capital but returning once each year to Jandaq to visit their relatives.

Introduction of the diesel pump for water lifting was less satisfactory in solving the problems of the Kavir villages than might have been expected. Some wells were brought into use and fed into the existing *qanat* channel, which was an ideal solution to the adoption of the new techniques. Importing the technology was expensive and uncertain. The severe winter of 1963 damaged many diesel sets since their owners were unaware of the need to keep them frost-free. Semi-deep and deep wells were often unsuccessful in finding adequate water to make irrigation worthwhile or, if successful, meant setting up cultivation at a distance from the established village area with all the attendant extra costs that entailed. Falling water tables were the inevitable result of intensive pumping in areas of shallow and limited aquifers, leading to diminished output from existing *qanats* and the drying up of shallow wells within the residential areas. The flexibility of the well and diesel pump, the encouragement of some of the authorities who had little regard for the *qanat*, and the eternal optimism of Iranian farmers that water could be discovered and profitably exploited, led to a great expansion in the use of the system alongside the *qanat* from the mid-1950s onwards.

5.3.3 The Mashhad basin, the Eastern sumps, the Zagros mountains and the mountains of the south east

Use of the *qanat* as a principal provider of water for irrigation was also a feature of the East Persian Highlands, their large endoreic sumps[19] and the Baluchestan mountain areas. Despite the belief in the underlying strength of the sub-surface water bodies in the Qa'enat,[20] the use of water of *qanat* origin in the east of the country was relatively small-scale, reflecting its high cost of development in conditions that were far from ideal for *qanat* construction. In particular, the rainfall in the eastern part of Iran was itself low in volume and poor in reliability with average rainfall less than 200mm a year over all but the upper slopes of the hills, and replenishment of aquifers was therefore slow and episodic.

Within the eastern region the *qanat* was used in parallel with local water diversion dams to conserve and divert water for agriculture known as *bandkari*, and was often supplemented by dryland cultivation (*daym*) on a wide scale during periods of heavy and well-timed spring rainfalls. The type of *qanat* varied considerably from place to place depending on the water table. Some were very long. The Aliabad *qanat* in Gonabad was four *farsakhs* in length (24 kilometres), for example, with a heavy discharge rated at 25 *juft-e gav*.[21] Though possibly not worse in this respect than other areas where *qanats* were used for irrigation, the east and especially Baluchestan tended to have a high proportion of workings that were difficult or dangerous. In the Juimand area of Khorasan an instance of this problem was the Delui *qanat* which was low-yielding and unrepaired because, it was reported by local farmers, no-one could be persuaded to undertake the work for fear of roof falls. *Moqannis* (diggers) shunned the village which was deteriorating in the early period of the land reform as water became more scarce. Because of its water problems, indeed, Delui was just the kind of village that was in part or whole handed over by landlords to the Land Reform Organization during the 1960s under the initial provisions of phase two (see next chapter).

Bandkari was an important form of water use throughout the eastern provinces. Temporary local barrages were thrown up across small streams and drainage channels to trap both water and silt, which could either be used *in situ* once the water level dropped so that a level and well watered triangular shape could be cultivated or be used to divert water along a contour or to fields lying below the dam. On occasions more permanent dams were used to divert water into *saddi qanats*, often short channels that led water from the area of the dam to cultivation below or adjacent to the storage system. Remarkably large areas of land were irrigated through the *bandkari* works, most spectacularly in the region of Zabol, where water diversion was practised and planting was sometimes undertaken in areas affected by light flooding.

The eastern highlands receive unreliable but at times heavy rainfall with cloud and mist clinging to the hills for several days. The area affected spreads westwards into the fringes of the Dasht-e Kavir and eastwards close to the Afghan frontier. Cultivation of rainfed lands depended on the incidence and volume of rainfall and was for the most part a gamble against the weather conditions. Some villages were entirely dependent on *daymi* cultivation. The village of Jangal in the Gaysur area of Khorasan comprised a dense population and was almost exclusively *daymi*. There was no *qanat* in the village and water was drawn from an *ab ambar* (storage chamber). In other villages the rainfall was monitored with great care since it affected the recharge of aquifers, influenced yields on irrigated lands and could make possible dryland grain cultivation. In years of good spring rainfall, of which 1964 was an excellent example, there was extensive dryland cultivation throughout the region. In Ferdaws, on the edge of the eastern hill lands, sowing of *daym* went into April in the hope of further late rains and a crop heavier than those in the irrigated gardens.[22]

In the poor environmental conditions prevailing in eastern Iran rural families made whatever use they could of slender resources. Over large areas of the region combinations of irrigation, diversion dams, floods, rainfall and dryland were exploited as and where possible in what was an insecure agricultural life. Indicative of the poverty of the agricultural base was the widespread resort to means of supplementing incomes. Carpet weaving was one response to the poor rainfall that affected much of the region during the early 1960s. There was a great expansion in weaving activities at that time since low wages made their products competitive with the traditional carpet weaving areas. Some small compensation was available for the adverse climatic conditions since saffron could be grown and picked in a number of the villages in the lower hills. The area was noted nationally for the quality of its saffron and moderate volumes of it were exported. For all attempts to diversify the sources of the regional economy, losses of people by emigration were high.

5.3.4 *The mountain rims of the Alborz and the Western Zagros*

The *qanat* did not entirely disappear as the plains and basins surrounding the deserts were replaced by the slopes and valleys of the high mountains of the Alborz and Zagros chains to the south, west and north. Underground channels were found in the valleys but in most cases played a variable role, dependent on whether villages had access to direct irrigation from rivers and streams or not. The latter provided water in the higher and narrower valleys and in the middle reaches where they traverse broad plains (*dashts*) to which water could be led. Classic techniques of irrigation in the high valleys were dams and particularly weirs. Few of the ancient dams survive in use. Temporary diversion weirs enabled water to be led around the valley sides gradually dropping down the contour in canals to command the cultivated terraced fields below. Weirs, canals and walled terraces, which were the integral elements of this system, required constant maintenance and heavy labour. Regional or national insecurity therefore had an immediate and adverse effect on the welfare of cultivation in these areas, which in the Zagros were often adjacent to what could be predatory pastoral nomadic populations.

Irrigation in the mountain areas was used in combination with rainfed cultivation where the terrain permitted. Large seasonal pastures represented an important agricultural resource used by nomadic pastoralists, and there was competition between the sedentary community and the pastoralists for land and water. The Lur tribes of the Khorramabad area studied by Bradshaw in the 1970s contained groups which were constantly at loggerheads over the relative rights of farmer and herder.[23] There was also a pattern of co-operation between the two kinds of groups brought about by ownership of agricultural land by the khans of pastoral nomads, so that the irrigated villages were accessible to herders on a limited basis in Bakhtiari areas. In the case of the Khamseh federation in the Zagros, there were powerful linkages between the pastoralists and the great landowners in Shiraz, while many Khamseh leaders invested their gains from good years in the herding economy in settled agriculture.[24] The khans of the Kuhgiluyeh held irrigated gardens in the hills above Daw Gombadan, though there was little evidence there of integration between the arable and herding sectors of the economy. In any case, the water and fodder resources of the villages were rarely adequate to prevent heavy losses of livestock in the pastoral economy during protracted periods of drought.

Cultivation in the high mountains was under constant threat not only from the depredations of the nomadic groups but also from migration. Long before migration became a problem in the lowlands, the upper areas of cultivation were subject to abandonment. Field terracing and even small settlements can be traced through their remains in the higher Alborz, suggesting a fairly protracted and possibly continuing withdrawal of the farming community in those isolated and hostile areas.

5.3.5 *The plains of Khuzestan and the northern Gulf coast*

Khuzestan province comprises a great area of unbroken plain made up of alluvium deposited over the northern extension of the Arabian platform. The drainage pattern is complex, made up of the River Karun in the north and north-east, the River Karkeh in the north-west and the river Jarrahi in the eastern quadrant. The irrigation history of the region is reasonably well documented and under the Sasanids there survived 'an elaborate system of barrages, tunnels, inverted syphons, lifting devices and canals',[25] which finally broke down under the Abbasids. There seems every reason to believe that the geographical extent of irrigation was never very great. Activity was closely defined by the courses of the main rivers and the principal service canals taken from them. Water channels changed their courses on several occasions, leaving cultivated areas literally high and dry. Yet, it seems that there was certainly far more activity in former periods than in the contemporary era in the years preceding the inception of the Khuzestan Water and Power Authority project in the 1960s. Settled irrigated agriculture before the project was confined to narrow 2-5km strips of land adjacent to the lower Karun in the vicinity of Abadan, Khorramshahr and Ahwaz, and to gardens in the other main oases that lay close to the river systems in their higher reaches, as at Dezful and Susangerd.

Irrigation practices in Khuzestan were varied and at times primitive. Water was lifted from the river by a variety of devices including wheels and buckets, which eventually gave way to pumping using powered mechanical systems. The large landowners in Khorramshahr all relied on water being pumped from the Karun for irrigation. Dependence on pumping enhanced the economic importance of the landlords who undertook the original large investment of the project and were responsible for its maintenance.

Cultivated gardens were often very rich in their variety and yields, which included fruits, vegetables and industrial crops. Date culture was important, especially in the gardens of the extreme south around Khosrawabad. The high water table throughout much of the south-west plains region enabled the sinking of shallow wells where water rights could be secured. Pastoralism, some shifting cultivation and tiny fragments of oasis culture were, however, the general pattern of agricultural exploitation, augmented by contract working of irrigated lands by outside, mainly Esfahani, groups.

In effect, the plains of Khuzestan were exploited very little. New works were undertaken under Reza Shah at Hamidiya on the Karkheh, at Behbehan on the Karun and at Khayrabad on the Sh'ur but all were small-scale, albeit useful, projects.[26] The region remained a challenge in so far as it appeared to have supported a sophisticated hydraulic civilization in the past and had an abundance of water resources and apparently vacant land available for development. The promise of the area was compelling and the government's adoption in the contemporary period of the regional development project, under the auspices of an American team with a background in the US water resource control and conservation experience of the Tennessee Valley Authority, seemed to offer a serious opportunity for the re-establishment of irrigated agriculture in the province. As we shall see, the early history of this programme was far from happy. Converting the potential into a productive reality proved beyond the government during the 1970s.

5.3.6 *The Caspian plains*

The northern plains are made up of the Caspian lowlands. Physical conditions are far from uniform but as regards agriculture the region is one that combines fertile soil with a generally high and reliable rainfall. This is supplemented by a variable but perennial flow of water in a series of north-flowing rivers that fall rapidly from the Alborz and its ancillary ranges.[27] Irrigation culture was remarkably varied. Offtake of water from the main rivers into diversion works and storage reservoirs has been important as a source of irrigation water for many centuries, the Sefid Rud in particular feeding major irrigation canals such as the one at Fumen and its many sub-branches. Modernization of the dam facilities at the Manjil gap in recent times extended the utility of the free flow irrigation facilities on the plain around Rasht. Elsewhere, water is lifted from shallow depths by pumping or *qanat*, the latter growing in importance in the Gorgan district to the east.[28]

Given the high rainfall of the region and the comparative abundance of surface and subterranean water supplies, the demand on the irrigation network was to make water supplies available at the optimum period for each crop during its growth cycle. Most crops on the plain were high water consumers — rice, citrus, cotton and tobacco — so that the water storage and distribution system assumed great importance and was managed with considerable sophistication. Water demand was so considerable at peak seasons that supplies were often under pressure, particularly if annual rainfall or snow melt in the southern catchment areas was below average. The extensive use of field terracing for rice culture added to the complexity of the valley and river fan gravity-fed irrigation patterns.[29] Perhaps more than any other region of the country, the northern plains saw a constantly improving irrigation system, spurred on by the mainly commercial, as opposed to subsistence, basis of farming. Constant modernization of irrigation facilities meant early adoption of mechanical water-lifting devices to augment traditional systems.

5.3.7 *Other irrigation regions*

So varied was the water resource development experience of the various agricultural communities in Iran that generalizations often fail to do full justice to the clever local and microscale adaptations utilized in irrigation. Combinations of river offtake, *qanat* flow, spring water exploitation, well systems and other irrigation works varied considerably as did the water storage structures such as ponds[30] at regional, village or farm levels that often went with them. Perhaps the common feature that recurs in all but a few irrigation regions is the *qanat*, a provider of water in Iran on a scale that has no equal elsewhere in the world.[31] The engineering mechanisms of the *qanat*, its perennial flow, and its negligible running costs other than for maintenance make it a powerful competitor as a provider of irrigation water. The agricultural and cultural rhythms that are associated with the *qanat* in all but a small area of Iran make it the *leitmotiv* of the country's irrigation ecology.

5.3.8 Dry-farming regions of the north-west and the Caspian plains

Dryland farming was rarely the sole interest of traditional villages, though exceptions, as at Jangal in Khorasan, did exist. Cereal and other dryland crops were an integral part of a balanced system of land use within village cultivations that were essentially based on an irrigated or semi-irrigated core area. The amount of dryland farmed each year depended on the volume of rainfall, its distribution through the season and the assessment of the aggregate soil moisture by each farmer. There was a high risk factor in the dry-farmed area which increased with the degree of unreliability of rainfall and the distance inland from the main mountain ranges. Yet, depending on the rainfall regime of the year in question, approximately half to two-thirds of the country's home-grown wheat came from dryland areas, by far the bulk (around 90 per cent) from the traditional system.

In Azarbayjan and parts of the Taurus and Zagros in the west of the country rainfall was reliable and heavy enough in most years to sustain a rich rainfed agriculture based mainly on cereal growing in the plains and foothills or orcharding in the narrower valleys and on the steeper hill slopes. None the less, even there the land-use pattern of the village community or the individual farmer would be diversified where possible by cultivation in areas where irrigation could be exploited, albeit on narrow terraces or in small gardens. In the principally subsistence farming systems of the western mountains it was important for farmers to have access to irrigated gardens which enabled the crop range to be extended beyond the limitations of rainfed production into vegetables, fruits and timber.

During the modern period tractor farming for cereals seemed to offer the means to revolutionize the dryland area on commercial farms. Large areas could be ploughed fairly rapidly using tractors and advanced mechanical equipment for ploughing, seeding and harvesting. Two quite separate options were open for mechanization, first, its introduction on the 3-5 million hectares already dry-farmed for wheat and barley and, second, the opening up of the extensive margin through reclamation of drylands under mechanized farming. To an extent the two options overlapped in so far as the conversion of former village *daym* ploughlands to commercial use took place as a result of the land reform as landowners sought to make their holdings ineligible for distribution though mechanization. Under the provisions of phase two of the land reform landowners could retain up to 500 hectares of barren (*ba'er*) land providing it was cultivated by mechanized means. Mechanization at Behdokht in the Gonabad region was an example of this trend in the early 1960s. Mechanization of dryland and other cereal harvesting in the traditional farmed area was also made possible in the 1960s through the availability of tractors and combine harvesters via contractors. Contractors, many of them from Gorgan and Dasht-e Gorgan,[32] travelled throughout Iran with their tractors and combines looking for work during the cereal harvest.

The most intensive mechanization affected reclaimed dryland. Okazaki analysed the development of the Gorgan area under capitalist wheat and later cotton farming in the period 1949-60[33] to establish a model for the process, which, as he perceptively recognized, was less than easily transferred elsewhere in the country. Despite this caveat, there was a modest degree of mechanization of commercial dryland cereal farming throughout the country and especially in the western provinces, that operated separately from the irrigated sector in some cases or on a scale that dwarfed the original irrigated centre of the enterprise.

The lack of success in the introduction of dryland cereals under mechanized farming techniques in the modern period can be explained in part by the disorganized scramble into the purchase of machinery during the late 1950s[34] and the failure of the technical and financial infrastructures for mechanization at that time. Certainly, multiplicity of makes of tractor, shortages of spare parts, abuse of the machinery through ignorance and inadequate commercial skills were major contributory factors which made the drive into agricultural mechanization slow and faltering. At the same time, the changing structure of agriculture had its adverse effects. Capitalist farmers felt constrained by the apparently unending stream of land reform and associated legislation which undermined the security of land ownership and added to the risks of investment in land reclamation and machinery. A significant number of owners of farm machinery hedged their positions after the land reform by providing tractors and combines mainly for contract use on other farms. In the main they refrained from developing land and farm businesses dedicated to dryland farming. The rapid growth in opportunities for rewarding investment in urban areas added to the disincentives for the allocation of private capital to agriculture in the period after 1968, and with particular effect after 1973. Government pricing policies for wheat ensured low returns for any investment made and were the death blow to mechanization of the so-called virgin lands under the 1965 legislation. This was considerably modified in subsequent years ostensibly to encourage the formation of large farm units on the *ba'er* and *mawat* (barren and dead) lands transferred to the private sector.

In the event, there was no revolution in the dryland sector. The largest increases in cereal cultivation during the period 1965-75 came on the irrigated lands.[35] Wheat and barley dryland acreages did grow but not at the rate needed to enable the population to be fed from domestic sources. The inability of the authorities to promote their policies of reclaiming the barren lands under mechanized dry farming also meant that diversification and specialization of cropping on the valuable but limited irrigated area of the country were inhibited by the need to keep a large portion of it, estimated at 1.5 million hectares in the mid-1970s, for low-value wheat growing. It is also remarkable that the modernization of dry farming was only partially effective outside the Gorgan area and its immediate environs.

Modernization was closely associated with mechanization since it was the introduction of the mechanical seed drill and soil-mulching cultivators that offered prospects for improvements in the short term over traditional practices. Where mechanization failed to penetrate there was also little likelihood of better cultivation techniques being applied or of innovations such as use of fertiliser, improved seeds or pest control being adopted. Similar conclusions do not apply to irrigated farming. There was some correlation between mechanization and the adoption of innovations in fertilizer or improved seeds even on irrigated farms as in Gilan and Mazandaran, but not to the almost exclusive degree found in the dryland sector. Cultivation in the irrigated fields was accompanied by a heavy use of labour over comparatively small areas, using agronomic techniques and watering skills that optimized the value of cheap hand labour. Dryland could not carry the costs of comparable labour inputs and there were no compensating skills to offset losses caused by broadcasting seeds by hand, lack of weeding and hand sickle harvesting. Physical limitations on the area that could be ploughed using oxen, especially in areas where rains were less reliable and sowing awaited indications of adequate soil moisture until the last possible moment, constrained the absolute surface area that could be sown under traditional methods in the short time period available. The one cheap and simple way to release a proven and substantial dryland potential was through mechanization and associated improvements that both increased the probability of good yields and enabled large areas to be sown rapidly.

5.4 Traditional irrigation and *qanat* culture

Estimates of the number of *qanats* vary enormously. Murray[36] quoted the Iranian Irrigation Department as suggesting a number between 32,000 and 40,000 *qanats* in 1950, while English[37] cited Ghahraman with a total of 37,500, of which only 21,000 were in good repair. Official statistics from the census of agriculture in 1352 (1973/74) suggested that there were 46,303 *qanats* still in use out of a total of 62,000, the majority in

Khorasan (10,929), the Tehran region (5,787), Kerman (4,938) and Esfahan (4,065).[38]

New *qanats* were being dug in the 1950s in the Kerman region at Negar[39] and in the early 1960s,[40] but few attempts were made to increase the length or number of *qanats* after that time in Kerman or elsewhere. In most areas the well and pump set became the vehicle for private water resource development and the large-scale dam and irrigation project played the equivalent role in the public sector. The *qanat* was thus apparently universally abandoned as a supplier of new water. Maintenance of established *qanats* deteriorated still further in the 1960s and 1970s so that the country crossed an important threshold, moving from an irrigation culture based on the *qanat* to new and often alien systems.

The reason for the appreciable, though gradual, change in the Iranian countryside arising from the demise of the *qanat* is the social and structural implications it carries for the communities which grew up around its water supplies. Ownership of *qanats* was often highly fragmented. Initial investors in a building programme might have been one person as in the case of the Javadieh *qanat*[41] or many hundreds as at the Qasabeh Shahr *qanat* at Gonabad.

Qasabeh Shahr *qanat* was reported by my informant as being owned by Astan-e Qods (the shrine of the Imam Reza at Mashhad) with 8×24-hour (*shabaneh ruz*) rotations, a private owner, Hajji Karimi, with one *shabaneh ruz* and a further *hazar* (thousand!) smaller holders of part shares. Since most *qanats* have been in existence for many years the ownership situation is now rarely so narrow as at Javadieh in the Kerman district. All those owners, farmers and their dependants who lived off the flow of the water from a *qanat* had a vital interest in ensuring its survival. The loss of a *qanat* would mean the collapse of the community it served, as was happening at Delui village in the Gonabad group in the 1960s. Reliance on the *qanat* by small and large landowners gave a powerful incentive for joint action in ensuring water supply, managing the irrigation cycle and preventing waste. It might be argued that for as long as the *qanat* remained the sole source of water it created and kept in being a strong element of social cohesion, even though this was expressed in many villages in oppressive relations between the more and less powerful participants in the irrigation system that it spawned. Agricultural and social command structures were tight and unforgiving for the ordinary peasants: participation was demanded from above and not sought in any co-operative way, but at least the fate of the whole rural community living off *qanat* water was to a large extent indivisible.

The *qanat* also clearly imposed its own influences on the morphology of villages and towns. The Oxford expedition to Persia in 1950 was quick to note of the channels of the *qanat*:[42]

> down they flow with the streams becoming smaller, dirtier and more insanitary; down to the poor part of the town, until finally the streams are no more: the clear stream of the qanat of the flowing jewels is then not even a muddy trickle.

Paul English codified similar responses in the selfsame area fifteen years later to a model based on the linear layout of villages, within which 'Household location with respect to the watercourse, therefore, reflects the social and economic status of its occupants'.[43] As in Kerman Province, so it was often elsewhere in regions of *qanat* water supply, with the richer owners of water monopolizing the head of the stream as it entered the settlement to draw off their water supply and irrigate their walled gardens before the intensifying process of pollution set in as the generally reducing volume of available water moved downstream.

The very flow line of a *qanat* and its main distributaries created pressures for villages to grow up in alignments along them. In its simplest form, the rich took the best of the water supply and the poor used its dregs and the social hierarchy in alluvial fan and valleyside villages literally dropped with the stream down the contour. More complex arrangements arose when several *qanats* served the same settlement or when relative or absolute changes took place in the flow of *qanats* within the same village. The principles at work, however, remained the same for each stream supply, as at Gonabad, for example, where three major *qanats* — Qasabeh Shahr, Aliabad and Deh-e Juimand — created their own distinct sub-hierarchies within the town.

The extension of villages into larger villages, towns and eventually cities was moulded to a degree by the system of channels and sub-channels feeding on the flow from the main water source. In the case of the *qanat*-fed areas of the country the linear and rectilinear pattern of the irrigation ditches designed to take their water to plots of more or less standard size laid down the geometry of the streets and the housing blocks that spread on to what had been irrigated farmland. The combined effects of the social gradient arising from the flow of streams from *qanats* noted by Smith and codified by English became a springboard for the work of Bonine,[44] who correlated the patterns of water distribution from the *qanats* and the streets to demonstrate how closely they matched in a wide range of settlements in central Iran.

De Planhol noted the 'powerful collective organization' that lay behind agrarian systems and rural settlement patterns in Iran[45] and linked the nature of *qanat* water supply with the morphology of the great Persian houses, courtyards and gardens.[46] Remarkably, he appeared not to have brought together his ideas on the communal command structure and the influence on it of the *qanat* or indeed other single-source irrigation structures that so markedly affected cultural phenomena such as gardens, pools and palaces. The inference of De Planhol's work was that the origin of the 'strict communal organization and thus overall control of the land'[47] arose from the survival of essentially pre-Islamic patterns within a strip farming system. In fact, as he partially recognized, strong social and economic structures were not the sole preserve of the strip-cultivated areas.[48] Strong social and economic cohesion was a common denominator of *all* settled irrigated agriculture throughout the country.[49] Communal economic interdependence was at its height where the survival of rural populations was not possible without irrigation and where there was a sole source of water supply. The *qanat* in most plateau villages represented the fullest expression of this situation, followed in sequence by villages with progressively increasing areas of rainfed land that diminished but rarely extinguished the central role in agricultural organization of the *qanat* or other irrigation water source. The correlation between the particular character of the *qanat*, including its slow rate of construction,[50] its high costs,[51] and the vital need for careful allocation of its output did more to explain the tightly-knit nature of economic and social organizations in much of Iran noted in the findings of the present author's fieldwork than any other single factor.

Throughout this study pains have been taken to refer in Iran to *irrigation cultures* rather than simply to mechanical structures and farming systems. The *qanat* for many centuries and in some areas to this day imposed a rhythm on life not just in the cultivated fields but in the *payabs*, the subterranean access points to the *qanats*, and other open air sites where the *jub* (ditch) was exploited during its traverse of the village residential area. It also influenced the flow of household water supply, the running of ornamental courtyard streams and pools, laundry activity and the flow of water through channels beside the village streets. Each *qanat* had its own cycle or *gardesh* that took it through different parts of the channels and fields of its service area on a regular rotation. In Gonabad, for example, the Aliabad *qanat* had a 12-day rotation, the Qasabeh Shahr *qanat* a 24-day rotation and the Deh-e Juimand *qanat* an 18-day rotation. The length of the cycle and the volume of water delivered by each *qanat* was crucial to the periodicity of water flow within the various sections of the canal system and therefore to the comfort and prosperity of those with shares in it, though the head water *jubs* running through village housing areas were either constantly fed or set to a closer time cycle than that used for field irrigation.

To live on the Aliabad *qanat* in Gonabad, for example, brought a rapid return of water in the *jub* after 12 days and a generous flow of water in what was rated as a 25 *juft-e gav qanat* (theoretically supporting 25 ox-powered plough teams over approximately 7.5 hectares). The main *qanat* of Gonabad, Qasabeh Shahr, offered a less easy rhythm with only half the rate of flow frequency through any part of its reticulation system as Aliabad. The lines of the *jubs* tended to set the pattern of the fields, the alignment of the tall plane trees that shaded them as they passed to and through the village and the orientation of the residential quarters. Ownership of shares in the *qanat* strongly influenced the wealth and status of families and individuals both by the agricultural or rent income it made possible and the location of the household up the water distribution channel. Wherever the *qanat* or the *jui*[52] water provision systems were the main sources of supply, these same features permeated all aspects of village and even town life.[53] An illustration of the flow of one of these, the British embassy *qanat*, is given by Overseas Consultants Inc.[54]

The mechanical and engineering ingenuity of the *qanat* has been examined in a number of studies.[55] Within the literature there has been no explicit recognition of the thoroughgoing influence of the *qanat* arising from its position at the heart of the agricultural system or its imprint on virtually every aspect of communal village life and the status in it of the individual.[56] Perhaps it was only with the gradual change away from the *qanat* to what are seen as cheaper and more instant privately or government-owned irrigation water supplies that the non-economic virtues of the *qanat* could be appreciated to the full. Oppressive or not, the economic and social organizations that arose around the *qanat* kept communities locked together on the land within a virtually closed hierarchical system. Those who could not survive within the system as owners, sharecroppers, craftsmen or labourers had little alternative than to leave the village. Social and religious pariahs faced the same fate or worse.[57] Over many centuries there must have been a wastage of people for a variety of economic and social reasons though, importantly, the main communities themselves remained in place unless the water supply itself failed.

In the contemporary period the rural population was assaulted in many ways by government intervention such as the land reform and the central planning of agriculture (see Chapters 6 and 7), changes in irrigation technology towards diesel pumping and dam reservoir supply, and the growth of the urban oil economy. Each of these elements attacked the stability of the established village institutions. The land reform removed the command structure from the villages, which, other than for a short time in the Yazd area, resulted in the collapse of the integrated control organization for the management of *qanats* in many villages. As an answer to the difficulties posed by a water supply for which there was no longer communal responsibility enforced by the larger landowners, the better-off and more active of the beneficiaries of the reform turned to well and diesel pumps to provide irrigation water. Not everywhere fortunately, but in many villages,[58] the most powerful social groups — the landlords and the entrepreneurial peasantry — lost their interests in communal water sources. Those *qanats* that were well built originally and had a high flow that could be maintained without frequent and expensive works survived with the character of their traditional structures weakened but intact. Those villages where new investment in diesel pumping made water available had the opportunity to remain in agriculture but lost the formative influences for social cohesion and nucleation of settlement formerly provided by the *qanat*.

Elsewhere over large areas of the plateau and the foothill regions, the *qanat* irrigation cultures were weakened considerably. Water supplies diminished at rates calculated at 3-4 per cent each year in Abuzaydabad and in the Dasht-e Kavir[59] for example, so that productivity and standards of living in villages affected in this way under the first phase of the land reform would have dropped considerably by the mid-1970s. Taken together with other forces exogenous to the countryside generated by the construction of a modern industrial and urban service economy and a later oil boom, the disintegration of social structures in many rural areas was far more dramatic than had been expected, largely because the authorities had no understanding of the organic nature of what we have called *qanat* culture. They were beginning the accelerated dissolution not simply of a set of traditional agricultural practices or an embarrassingly oppressive landlord regime but of a fairly comprehensively enveloping way of life built around a communal but eccentric water supply system.

5.5 The survival of the *qanat*

One of the greatest virtues of the *qanat* is its generally long life cycle. Once completed, flowing and given adequate maintenance, it will run day and night and throughout the seasons without the application of motive power other than that provided by gravity flow. Changes in government policies do not immediately affect water provision from *qanats*, though encouragement of pumped wells has an adverse impact if pursued over an extended period of time. It will remain, therefore, an option for the authorities at least to maintain existing *qanats* if not to revert to the construction of new ones. Certainly, in the 1980s under the revolutionary government there were some indications that *qanats* were returning to official favour, suggesting that the balance of argument on the issue of their abandonment had shifted.[60]

The advantages of *qanats* are considerable and often understated in the literature. A balance sheet of factors for and against their use was prepared for the Plan Organization in 1949 by Overseas Consultants Inc. (OCI) and by Henri Goblot in 1979.[61] The former showed two positive and eleven negative features. The only favourable aspects of the *qanat* according to OCI were its low costs of maintenance and the negligible foreign-exchange requirement for its initial construction. There is much more to the *qanat* than OCI stated, though their antipathetic views have tended to reflect and reinforce approaches to the *qanat* by the official planning authorities throughout the modern period.

Among the most useful characteristics of the qanat is its benign influence on the water table. Its withdrawal rate is naturally adjusted to water availability and is altogether different from powerful extractions like those of motor pumping in wells that cause spreading and deepening cones of depression in the aquifer. Pollution of aquifers by *qanats* through salt and brackish water intrusions is unknown, though these problems became common in many districts adjacent to the central deserts in the 1960s and 1970s as well drilling projects went ahead.[62] Provided the *qanats* are protected by traditional practices and legislation to defend their *harim* (the protected area around the well within a variable perimeter depending on the nature of the water table and soil structure) against other well digging, they remain safe against loss of their supply short of natural catastrophe of roof fall, earthquake, flooding or regional desiccation and prolonged drought. In normal conditions, too, water supply does not cut off completely, as can happen as a result of mechanical failure with a motor pump, which is a very important consideration in villages distant from sources of spare parts and mechanical expertise.

The *qanat* was a transporter of water from areas of water availability to areas of water need and offered benefits of a special kind in an arid country such as Iran. Most *qanats* ran for long distances undergound and in consequence carried water in a conduit where evaporation losses were kept extremely low vis-à-vis channels open to the air. Seepage through the beds of the *qanats* was, of course, a feature of most water channels, the rate of loss varying with the porosity of the soil. Few water courses were fully lined except where they traversed soft sands or other difficult terrain where the *qanat* makers, the *moqannis*, would use baked clay hoops as linings.[63] Occasionally *qanat* beds were sealed in areas particularly susceptible to high levels of seepage through adding fine clay to the open stream. The clay would settle in the coarser sands of the bed and provide a form of water-resistant puddling.[64]

Two particular advantages of the *qanat* stood out during the 1980s. First, *qanats* are indigenous in all senses. Survey, construction and maintenance, together with supply of equipment and personnel, can all be done with Iranian and mainly rurally-based resources. *Qanats* carry with them absolutely no external dependences nor, as OCI correctly pointed out, foreign-exchange costs, a matter that took on some significance during the harsh war years after 1980 and was of particular value during the years of national financial crisis after 1984. Second, the revolutionary authorities were eager, especially in the immediate aftermath of the fall of the shah, to re-establish the religious, economic and cultural values that it was felt the former regime had destroyed. The keepers of these virtues were felt to be the rural towns and the traditional villages, the small-scale enterprises and the farms.[65] The initial momentum for positive renovation of the country's agricultural stock was lost soon after the revolution and not least as a result of the government's preoccupation with the war against Iraq that began in September 1980. Thereafter, the protection of the *qanat* and its role in irrigation arose from the slow rates of development in alternative methods of water lifting. The import of pumps and spares became increasingly difficult and unreliable as bank credits for imports were given with growing reluctance after 1983. At that stage there was an equilibrium in the comparative roles of *qanat* and pumped well.

Favourable treatment for traditional systems of farming and local technology solutions to problems such as the *qanat* did not overcome some of the inherent constraints of its use as a principal source of irrigation water. Initial capital costs for *qanats*, albeit denominated in Iranian rials, remained much higher than for wells.[66] Water flow from *qanats* is maintained for 24 hours every day of the year. In many ways, illogically, Iranians tend to look on the water that runs through the system during the winter when irrigation is not being used as wasted.[67] A fairer view of the problem is that matching supply to demand in a *qanat* system was difficult and mainly impossible in the short term. Techniques of improving the flow from the mother well were available by extending the tunnel deeper into the water table and establishing a new mother well or *madar chah* in a process known as *pish kar kani* or creating lateral feeder galleries. Supplies could be augmented by feeding water from new short *qanats* into the existing stream. Meanwhile, underground tunnelling was slow. It was estimated that the average time to build a *qanat* ranged between two and seven years,[68] though much longer periods were recorded, such as 17 years for construction of the Javadieh and 27 for the Hojjatabad *qanat* in Kerman cited by English.[69] In Behdohkt, the mayor told me in 1964 of a *qanat* that had been 45 years in the making.

It will be apparent that the greatest difficulty in the use of the *qanat* as a mass instrument of water provision remained its inflexibility. In a slow growing population where food supply was increasing at a moderate rate this difficulty might not be acute. In Iran, with its rapid increase in population and severe dependence on foreign food supply, more urgency would seem to be required than can be met by the *qanat*, unless the rainfed sector could be strongly developed for cereals, leaving the irrigated lands to higher value cropping in which the *qanat* could play a useful part. Up to the present time, however, no balanced and rational policy of this kind towards the role of the *qanat* has emerged, and its future and that of the residual areas of the *qanat* culture are at best uncertain and at worst at great risk of extinction.

5.6 Modern irrigation

As noted in Chapter 2, the need for more irrigation in a rainfall regime that is inadequate to support dryland farming over much of the country and is susceptible to radical variations in amount, incidence and periodicity from year to year and place to place is considerable and growing. The availability of water with which to provide new sources of irrigation is less well demonstrated, since the climatic and hydrographic data required to calculate the situation are generally poor. In theory, the country has the opportunity to exploit a large amount of water that would otherwise run to waste.[70] Two means are open to limit the water loss. First, the catchment basins of the great river systems within the country could be made less liable to rapid and dangerous run-off by means of afforestation, improved cultivation practices, control of grazing and water spreading to improve additions to groundwater reservoirs. Second, river and stream flow could be regulated and stored so that water which ran to the internal salt basins or the external seas could be held back.

With an exceptional degree of self-confidence and panache the government began to implement programmes for both conservation and storage on a catchment-wide basis in the Khuzestan Water and Power Authority scheme in Khuzestan in the 1960s. This was based on studies recommended by Overseas Consultants Inc. in 1949,[71] financed with IBRD assistance, and implemented from 1957 with Dutch and US (Development and Resources Corporation) technical support.[72]

There was a high degree of achievement in dam building within the very large project area in south-western Iran. The first concrete was poured for the initial dam above Andimeshk in 1961 and the unit was completed in 1963.[73] Other dam projects in the proposed development area[74] were eventually put in hand so that much of the civil engineering works associated with the storage of water were in place by 1979. Unfortunately the concept of total environmental management borrowed in the initial project from the Tennesee Valley Authority programme was lost. Tree planting attained only the cosmetic level and the vast upper reaches of the tributaries of the Karun were untreated. When serious flooding occurred, silt flow into the reservoir was reported to be heavy, several years of life of the reservoir being lost after each sustained torrent.

Agricultural use of the stored waters of the Dez project came very slowly. A 12,000 hectare sugar-cane growing area around Haft Tappeh in the south of the development area was brought into use in the year 1961/62 together with a 200-hectare experimental farm. The larger zone, made up of a gross 145,000 hectares, was partially reclaimed piecemeal under the regulations for agribusiness managed by the Ministry of Water and Power. Five companies, all but one with foreign participation, operated units in the area, the first farm being established in 1969/70 by Iran-American. Their performance was poor both in the amount of land they physically reclaimed through levelling — a mere 30 per cent — and in the value of their output.[75] Hydroelectric power was drawn from the turbines at the main dam site, but the environmental and agricultural impacts of the project were deeply disappointing.[76]

Meanwhile, modern developments based on storage dams of a less spectacular kind were begun at twenty other sites in the period 1955-78, seven of which remained incomplete as late as 1986. In all, the reservoir dams were supposed to provide irrigation supplies to some 745,500 hectares of land throughout the country. If the deflator factor devised by Nattagh for Dez to reduce gross hectares to reclaimed cultivated hectares[77] at 30 per cent is applied across the board, even though this might err on the generous side, no more than 225,000 hectares would have come fully under new irrigation regimes during the five development plan periods ending in 1978.

The largest contribution to expansion of the irrigated area in the modern period must be attributed to well development, mainly by private entities. In 1950 Murray estimated that traditional wells in which water was raised using animal power provided approximately 0.5 per cent of all irrigation water supplies for some 8,000 hectares.[78] Tube wells yielded some

0.2 per cent of all water supplies feeding 3,000 hectares. At that time *qanats* and springs made up some 60 per cent of all supplies, serving 1,200,000 hectares. Increasing use of well water supplies was constrained by the authorities on the grounds that unrestricted well drilling would needlessly put aquifers at risk if made in advance of a national survey of subterranean water resources. None the less, there was a great deal of drilling, some illegal. Official sources estimated that approximately 3,000 million cubic metres of water were extracted in this way by 1968.[79] The Fourth Plan provided for construction of 1,500 deep and 3,500 shallow wells,[80] so that by the mid-1970s the volume of water lifted by tube wells rose to some 8,000 million cubic metres against 9,000 million by *qanat*.[81] There was little evidence available to indicate the success ratio of wells drilled and the drop-out rate of wells that ran out of sweet water or suffered from mechanical or commercial failure.[82]

Well systems created very different social environments from *qanats*, even where apparently traditional economic and human relations were practised. Water generally belonged to one man or a small group of individuals, often members of the same family. *Jubs* leading water to the irrigated fields were quite separate from the village water supply, so that the patterns of distribution of housing and gardens in the *qanat*-fed village were rarely reproduced in the new settlements. Similar considerations applied to the layout of the fields, since pumped water could be fed from raised tanks in a variety of directions and not simply down-slope from the source as in the case of *qanats*.[83] Ownership of land was in many cases uncomplicated by ancient rights and customs, water distribution was adjusted to irrigation needs rather than to a cycle demanded by the complex rotation of *qanat* water, and field shapes were governed by the immediate needs of cultivation practices and the type of crop being grown as opposed to traditional practices. In several of the pump settlements that were visited by the author in a variety of areas of the country, it was noticeable that housing was only loosely grouped and often settlers or workers would set their houses at a distance from each other, usually with each house located near a pump unit and water storage tank. Traditional street and house layouts within a nucleated core were characteristic only where well schemes were brought into established rural communities or where conditions of insecurity were felt to prevail, as in the Afghan frontier regions east of Gonabad. The modern *shahraks* or rural towns set up on the Dez scheme[84] were modelled on Western concepts and were entirely divorced from traditional settlement forms in what was in any case a region rarely involved with *qanat* water provision.

5.7 Conclusions

Water lies at the heart of the Iranian agricultural dilemma. The traditional systems of water provision were adjusted to the harsh and hostile environmental conditions within a technology that Iranian rural peoples had developed to a fine art. The various climatic, topographical and geological characteristics of the country helped, with human factors, to create a distinctive set of irrigation regions. Yet, for all the regional diversity, the *qanat* particularly and the *jui* as a surrogate from place to place created a common strand within irrigation mechanisms. The great heartland of the plateau and the surrounding mountain basins was the home of the *qanat*, which was rarely entirely missing elsewhere in the country. It has been argued in this chapter that the *qanat* and the culture that grew up around it comprised a powerfully integrated communal system dominated by an internal and authoritarian command structure based on the long-established principle that societies that failed to maintain their water source were doomed to poverty, dependence and ultimate extinction.

The very nature of *qanat* water supplies also set a rhythm to life in the villages. Channels that led water from the underground source through the village powerfully influenced the patterns of streets and the location of different qualities of housing within each settlement. Downstream of the village centre, the reticulation of water across the irrigated fields took on a geometry that arose from the adjustment of agricultural land-use areas to the seasonal limitations of the generally inflexible *qanat* water supply. This set of responses is not unique in itself but it is unusual in its scale and character, as W.B. Fisher pointed out.[85] All societies that rely on irrigation for agricultural survival and are dependent on a sole or limited water supply of this kind will tend to exhibit similar cohesion.[86] What made Iran somewhat unusual was the longevity of the system, though not necessarily of individual *qanats*. The system, as De Planhol notes, was indisputably pre-Islamic,[87] and its internal strengths were that each unit could operate as an agriculturally autonomous entity, and, equally, the destruction of one or several *qanats* had little effect on the rest of them.[88]

The primacy of *qanat* culture was challenged by a number of related phenomena, almost all of which arose in the twentieth century. Growth of population became more rapid, stimulating an increase in demand for food which could not be met easily by improvements in cultivation practices nor by expansion of *qanat* water supply, which was unequal to adapting to changes in the short term. Development of the country's oil industry created two new conditions that affected the position of the *qanat*. Internal distribution of oil products and the spread of the internal combustion engine not only opened up the countryside to external influences but specifically made possible the introduction of the diesel pump for water lifting as a direct competitor to the *qanat*. Later, the oil industry set in motion a host of indirect effects on the economy that disadvantaged traditional rural society vis-à-vis urban areas and considerably changed the economic structure of the country to the loss of agriculture as a whole. The sequence of the choice for those in the villages that enabled them, first, in a specific sense to opt out of the communal *qanat* system into individual or small group ownership of water-lifting devices and, second, more generally to find better paid and less uncongenial work than agriculture, normally in the towns, helped to undermine the basis of economic and social strength in the villages. In undermining village cultures by these varied means, the oil industry was responsible for a perceptibly increasing rate of erosion of what had been the most abiding and influential socio-cultural components of the country.

On the other side of the scales, it is evident that use of funds — mainly earned through oil exports — in investments such as large-scale multipurpose dam projects failed in the years 1955-79 to provide a successful alternative means of augmenting material output from irrigated agriculture. Private and government pump schemes did rather better, but seemed to carry a high loss rate and caused damage to the *qanats* in some areas through drawing down the water table below their intake levels. Official concern with preserving groundwater must be applauded. Well developments were controlled, albeit with less than total efficiency, throughout the modern period and in most regions of Iran total anarchy in water use, such as prevailed in other arid states of the region, was avoided.[89] This at best passive approach was inadequate to save the *qanat*, not least since the authorities failed to act against new pumped wells which shared local aquifers with *qanats* rather than segregate their respective areas of withdrawal.

It is argued here that a distinctive irrigation culture evolved around the idiosyncrasies of the *qanat* as a water source and became in many ways a formative influence on significant parts of the Iranian inheritance. It must also be concluded that this aspect of Iranian life will be lost in most, if not all, villages. At the general level 'village' communalism will become a second-hand experience, as the source of it declines and as the majority of Iranians grow up in urban environments.[90] The disintegration of the *qanat*-based villages as the use of alternative water-lifting devices spreads will slowly ensure that even many of those born and raised in villages will live outside the *qanat* communities. If it is accepted that the *qanat* provided more than just water and exerted its own special influences on the way village settlements were structured architecturally, morphologically and socially in the manner that this chapter suggests, the passing of the *qanat* and the special kinds of villages it gave rise to has already altered and will further alter more than just the face of irrigated production from the agricultural sector.

6 Change and Development in the Countryside: Land Reform and Centralization

6.1 The move to land reform

The forces working for change in the countryside in the wake of the shah's restoration to the throne by the Western powers in 1953 were few. Significant investments were made in agricultural and irrigation projects under the terms of the Second Plan (1954-61) but they were by their very nature geographically highly concentrated in the new dam/water storage schemes put in hand on the Karaj, Dez and Sefid Rud rivers. Elsewhere the efforts of the state-run organizations responsible for improvements in agriculture were inhibited by the lack of a co-ordinating planning framework, shortages of personnel, an absence of adequate information on conditions in the countryside, and budgetary constraints.

Continuation of the crown lands distribution programme pressed ahead, though, as we saw in Chapter 4, in the teeth of mounting opposition from a number of landlord interests.[1] The plot by plot distribution of the 800,000 hectares controlled by the Pahlavi Foundation kept alive the uncertainties of the landowners and was an example of which peasants cultivating areas outside the royal domain were aware. Added impetus was given to the allocation of lands to those that worked on them in 1955 and 1956 when the decrees were issued reinstating elements of the Full Powers legislation set up under Dr Mosaddeq as it affected village councils. The change was not to be precipitate, however, since the regulations manual for its implementation was not made available until 1957 and intensive activity began only in 1958. Although generally prosecuted in good faith at the farm level, the need for a land survey in each location where distribution was undertaken meant that progress was painfully slow.

The activities of the American aid agencies grew rapidly under the programmes of the Ford Foundation and the United States Operations Mission, while the United Nations was also present through FAO, UNESCO and other agencies. Yet the broad area of rural Iran was untouched by the institutions acting in support of agriculture and even in those villages affected by development schemes there was rivalry and a failure by government and other departments to integrate their work.[2] Most villages were rarely touched by the central government except in the domain of security through the gendarmerie. Poverty of resources in maps, cadastral surveys and taxation documents meant that a number of villages were altogether unknown to the authorities.[3] Certainly, those sparsely settled areas of the country distant from the major urban centres or in isolated desert or mountain regions were regarded by most government officials as beyond the pale of their interest. The greater the distance from Tehran, the greater the lack of involvement by the authorities and the more the reluctance of civil servants to travel to or spend time there.

Such elitist attitudes continued to prevail and were reflected in the low quality of much of the administration in rural areas of the country and resulted in policies being formulated for agriculture which were devoid of understanding of the many and acute differences in land, water or agronomic practice between Iran's varied geographical areas.[4] A notable exception was the Agricultural Development Bank. Under the terms of various separate legislative acts, including the 1951 *khaleseh* regulations, the Full Powers Act and the 1956 Village Councils Act, the Bank began to penetrate much of the countryside and come to terms with both the nationwide and regional problems of the farming communities.

Although the rural areas were largely left to their own devices by the development agencies during the mid-1950s and serious interest in agricultural production or social conditions was rarely expressed, there was change in the wind by the end of the decade. The motivations for the government's new approach to agriculture that began to be felt in 1959/60 have been variously attributed by informed observers. Professor Lambton suggested that the shah and those Iranians with political influence were strongly affected by the need to keep American and other foreign goodwill so that financial and military support from outside should not be put in jeopardy.[5] It was sensed in Tehran or possibly forced upon the government by the US ambassador and other prominent Americans that the Kennedy Administration could not support a backward regime of the sort current in Iran. The government therefore turned in haste to a more radical reform of the agrarian structure with which to please its principal foreign ally.

There were also what proved to be significant internal Iranian pressures pushing for reform. Dr Hasan Arsanjani had pressed for changes in land holding in the 1940s and early 1950s.[6] There were others who in the late 1950s were publicly attacking the landlord system, presumably with official support. The unmitigated poverty of the peasants subsisting on as little as 10,000 rials a year,[7] the inequalities of the tax system and the unjustified dominant position of the landlords within the rural economy were all roundly attacked.

There were only minor demands for land reform from the political parties and dissidents of the left, whose role must be seen as peripheral at best. Most Iranian socialists and marxists saw their constituencies in the urban rather than rural areas. The Tudeh Party in particular was widely seen as uninterested in the land problem[8] and there had been little done to alarm the landlords during the brief Soviet-supported independent republics in the north-west after the Second World War. National Front concern with agricultural change had been demonstrated to be strictly within the confines of modifications of the landowning *status quo* during the government of Dr Mosaddeq. It is true that a number of individual Iranian political thinkers put forward ideas for the socialist transformation of the rural sector but only as comparatively minor parts of their general writings. Such views were not influential except on small numbers of intellectuals, many of them disciples after rather than before the promulgation of the Iranian land reform laws.[9]

The part played by the shah in bringing about the land reform was plagued by controversy from the outset. Various biographies indicate that he himself was distressed by the poverty of the peasantry, displeased with the implacable pursuit of self-interest by the landlords, and angered by their failure to heed the warning signs that the government would no longer accept the maintenance of an agrarian system that so depressed the living standards of the mass of the peasants and precluded the introduction of greater levels of agricultural productivity.[10]

Explanations for the shah's stance were varied. Professor Lambton was convinced that he and his ministers acted mainly with a foreign audience in view.[11] Interpretations by writers of the left suggest that the shah was concerned to consolidate bureaucratic power at the centre and, by implication, to negate any remaining effects of the constitutional movement of 1906.[12] There was opportunity for imposing a marxist interpretation on the events of the land reform. The reform could be interpreted as a paternalistically imposed change that replaced the traditional landowning class by middlemen operators, the *gavbands*. It was an argument of some merit that the *gavband* had acted as an intermediary between the former landlord and the cultivating peasant, providing one of the inputs to the agricultural process, normally oxen for ploughing. As noted later, the *gavband* was to be found in only limited areas of the country and in certain kinds of farming contracts. Despite this severe limitation, it was suggested that during the land distribution programme the *gavbands* in several areas of the plateau took possession of land instead of the former cultivating peasants. Using this evidence, Halliday, for example, took the view that a capitalist class system was erected in place of the former landlord-peasant structure.[13] Presumably, this socio-political change was an inexorable part of the process of revolution in which the shah was a player for a short-term gain that would ultimately bring about his own downfall as a result of the politicization of what would become an increasingly disaffected peasantry.[14]

Opinions of those on the left concerning the shah and his involvement in land reform are more available than firm objective evidence on the matter. Such partisan appraisals none the less tend to contain elements of the truth, though they also purport to represent for the most part overall views of the reform, while their utility lies in isolating one facet of activity at one period, often without attachment to a specific geographical base. Bearing in mind that the agrarian reform programme ran from 1960 and was still evolving during the 1970s, encompassing a variety of different laws, regulations, economic situations, regional environments and political conditions, it is not difficult to understand why informed observers have offered so wide a range of explanations for the shah's actions and attitudes.

To offset the notion that the reform was simply a response to foreign pressures highlighted by the election of John F. Kennedy to the Presidency in 1960, it must be pointed out that the shah had been a consistent advocate of modernization and development since his early clashes with the fifteenth *majles*.[15] His distribution of the crown estates, his attempt to encourage other landlords to share out their lands among the peasantry, and his apparent conviction of the solid virtues of cultivating the support of the farmers are a fair record of concern unmatched in ideal by any of Iran's political parties. The shah's belief that the landlords were a political obstacle to him through their strength in the *majles* and the administration and that they needed to be removed was not inconsistent with his dislike of them as a group which stood between him and the achievement of a more productive agriculture. That the shah was eager to press through reform to please those on whom he depended in the USA was matched by a desire to consolidate an internal political bond between himself and the peasantry which could be most readily achieved by elimination of those large and influential landlord families aligned against his interests and the creation of a policy that directly benefited the cultivating peasants.

It seems probable, too, that deteriorating domestic economic conditions at the end of the 1950s led the shah to adopt the reform programme. He realized that the financial support of the US Government was crucial,[16] but that such aid would be forthcoming from an increasingly critical administration in Washington only if his domestic policies were seen to be liberal. He also needed the reform to divert attention from the swingeing effects of economic recession at home. Meanwhile, the reform programme offered a low-cost mechanism with which to save the country from losing all the momentum for economic advantage gained during the years 1954-59.

Above all, the shah had to find the means to keep his regime in existence through yet another period of crisis. He had shown consummate political adroitness in surviving the abdication of 1941, the domestic crisis during the Anglo-Iranian oil dispute, and his flight from Iran in August 1953. In comparison, his problems in 1959-61 were less taxing in so far as he appeared at the time to be seeking a new form of legitimacy with which to lay the basis for long-term survival rather than overcoming a short-term crisis. The new legitimacy which emerged perhaps piecemeal was entirely in keeping with his own predilections for modernization of the state and development of the economy with as little interference as possible from the *majles*, the informal political opposition or the forces of economic conservatism.

It was ironic, as Mahdavy so correctly foresaw in 1965,[17] that the shah's route to 'modernization' through land reform should have so thoroughly alienated the *'ulama* and many of the educated Iranian elite from both the regime, which was predictable, and the United States. This was a high cost to pay. But it is difficult to believe that the shah could have escaped a confrontation with the political opposition, including the National Front, and the *'ulama* on the constitutional issues that arose with increasing acuteness from 1960 onwards, even if he had compromised on land reform. In the event, land reform among other items became an area chosen for conflict by the shah and his opponents. In so far as the shah appeared as the ally of the United States, the latter was inescapably destined to take over the British mantle as the 'southern power' counterposed to the USSR in the north. Among the several ways in which the US presence in Iran was different from that of the British were the total linkage between the Americans and the personal rule of the shah — a bilateral arrangement that the British had never permitted themselves — and the American association with the general process of modernization and militarization of Iran, with which the British had never historically had a parallel stance.

6.2 The land reform

The land reform proper began in 1960. Those preparing the legislation appeared to be more concerned that a reform law should be under discussion, especially during the shah's visit to the United States, than with the formulation of a consistent and well-designed bill.[18] In retrospect, the 1960 land reform bill, which provided for a universal ceiling of 400 hectares of irrigated land or 800 hectares of dryland, seemed to be little more than a sacrificial and tentative first move.[19] The reform was denounced in principle by the *'ulama* as an infringement of Islamic Law and the bill was modified during its passage through the *majles* into an anodyne measure of little threat to the real strength of the landed families and ultimately left without authorized regulations for its implementation. Its fundamental importance lay in the passage of the Land Reform Act through the *majles*, since this act was used later as the legal basis for more effective legislation to be given retrospective force despite the absence at that stage of the national assembly.

The land reform programme emerged from a decree amending the land reform law of 1960 signed by the Prime Minister and some of his ministers on 9 January 1962, whereby a maximum limit of one whole village was imposed on land holdings. Landowners possessing more than the maximum were required to dispose of excess land to the state. The terms of expropriation were lenient, with the owners permitted to retain a village of their choice together with all lands under mechanized arable farming, orchards, plantations, and woodland for which they held undisputed rights to land and water. Compensation, often on a generous scale, was payable to land holders affected by the law.

Lands above the ceilings on individual ownership were to be

purchased by the government and transferred, together with their water supply, to the cultivators of the village concerned. Payment was to be over a fifteen-year period, and beneficiaries were obliged to repay the purchase price from the ex-landlord plus a service charge not exceeding 10 per cent of it. Transactions were to be conducted through a Land Reform Organization established by the same law. A Land Reform Council was also set up under a chairman appointed by the Minister of Agriculture, on which the head of the Land Reform Organization and four senior members of the ministry were represented.

At its inception, the Land Reform Organization foresaw that some 13,907 villages (consisting of 3,788 entire villages and 10,116 part villages) would be affected by land transfers under the terms of the 1962 decree.[20] In order to spread the effects of the reform to the remaining villages untouched by land distribution, share-cropping contracts were improved by an overall increase of 5 per cent in the peasants' share on irrigated land and 10 per cent on dryland. These various measures became known as phase one of the reform.[21] Implementation began once regulations were issued by the authorities (see below). A variety of publications in Persian were also available to explain the reform to the population.[22]

After operation for one year the Land Reform Decree was broadened in scope by the Additional Articles of 17 January 1963. These stipulated that the maximum limit of individual holdings was to be 100-120 hectares in the dryland areas of the south, varying through a graded range to 30 hectares in the irrigated rice-growing areas of northern Iran. Owners with more than the new limit for their areas were to dispose of their excess holdings to cultivating share-croppers by lease, sale, division or by forming a joint company with their former peasant cultivators. Provision was also made for landowners to buy land from those peasants willing to sell, an indication of a considerable change from phase one in the approach to the reform. It was expected at the time that the January 1963 legislation would bring the total number of villages affected in whole or in part by the reform programme to approximately 33,000, between half to two thirds of all villages. Phase two of the reform withdrew rights of landlords to retain orchards, plantations and woodland above the stipulated maximum holding, though mechanized farmland of up to 500 hectares could be kept. Compensation was provided for all lands forfeited under the law. The draft regulations pursuant to the Additional Articles, passed on 25 July 1964, introduced significant changes including, for example, omission of the definition in hectares of the amount of land that might be retained.

The design and implementation of the land reform was largely under the control of Dr Arsanjani, the Minister of Agriculture during the politically crucial period 1961-63. His aims were supported by and large by the Prime Minister, Dr Amini, who was himself a large landowner in the north of the country. Dr Arsanjani had experience of the workings of the Agricultural Bank, and as a lawyer and journalist had a long-standing dedication to land reform. His political motivations were considerable, and for this and other reasons the reform failed to win the clear support of most of his cabinet colleagues or indeed of the shah. Cottam has argued with some merit that Arsanjani had a specific vision of transferring power from the traditional elite to the peasantry as a means of channelling political change into evolutionary rather than revolutionary paths.[23] The shah appeared in the first instance content to acquiesce in Arsanjani's mobilization of popular support for the reform but later became suspicious of the minister's growing personal following. The nationalists and the left-wing opposition, as might be expected, found Arsanjani's activities to be an anathema. He was roundly criticized as an agent of the shah and his land reform proposals represented as a sham.[24] The poor showing of the later phases of the reform, the short period during which he was in power and the deliberate understating of his role by the shah's regime have unreasonably diminished his reputation.[25]

Dr Arsanjani was an adept and purposeful manager of the land reform programme. He created a new arm of the Ministry of Agriculture composed of committed engineers and administrators as the Land Reform Organization, and this permitted him to implement the programme without reliance on the established bureaucratic processes in which delay and obstruction would certainly have been met. Under Arsanjani's direction the reform programme was divided into two phases, the first affecting only the owners of large estates and the second all other land holdings except those belonging to peasant proprietors. In this way the Land Reform Organization was able to concentrate its slender human and material resources on redistribution of a comparatively few large estates in the first instance and later, after a two-year period of expansion and field experience, begin the purchase and sale of the mass of medium-sized holdings. Politically, too, the separate assaults on the large and medium landowners was expected to facilitate the reform by dividing the opposition to it. In the event, the reform was introduced without bloodshed and unrest in the countryside. Acts of violence against land reform in the period 1962-64 were largely confined to urban areas, particularly Tehran.

During the early period of the reform, Dr Arsanjani exploited to the full the opportunities provided by the national and provincial radio networks to propagate the ideals and aims of the reform and also to discredit the landlords. The coincidence of the reform and the spread at that time of cheap radio sets throughout the villages strengthened the broadcast of Land Reform Organization materials, some of them virulently anti-landowner. At the same time, Arsanjani took pains to gather around him a small but reformist group of agricultural engineers who had allegiance both to himself and to the reform programme. For the most part, the group did much to gain the confidence of the peasants as agents of their interests against the landlords and in some cases against the bureaucratic establishment at large. During the period January 1962 to September 1963 the group completed the first phase of the land reform, except in areas such as Kerman where complex land and water ownership delayed settlement.

Dr Arsanjani resigned as Minister of Agriculture in early 1963, after which the terms of the second phase of the reform were considerably modified. The original proposals had foreseen the emergence of small peasant proprietor farms throughout the country. Under new legislation prepared after Arsanjani's departure by the Mansur Government in May and June 1964, the second phase permitted landlords to remain in their villages providing that they established a 'joint company' with the share-croppers, rented their lands to the share-croppers on improved terms, divided their lands with the peasants in proportion to their contributions of factors of production (shares out of five), or bought out their peasants' rights (*haqq-e risheh*). Within the legislative framework set up through the modified Land Reform Law, government policy from the spring of 1964 worked towards the creation of a two-sector agricultural structure, with peasant proprietorship and tenancy on the one hand and large-scale mechanized agriculture and medium-scale commercial farming on the other.

In practice, the land reform had rather different implications at various times following the January 1962 decree. Only for a short period was the reform a land redistribution programme. Restricted and revised leasing of land was ultimately the main effect under phase two. Alongside changes in land tenure there were provisions for new agricultural support services, ranging from credit and co-operative organizations to literacy and health facilities that penetrated to most villages, the successes of which were real in the early period 1963-66 but thereafter less so.[26]

The government officially declared the second phase of the land reform completed in January 1967, five years after the inception of the overall programme. In fact, there were small pockets of land where a final settlement had not been possible as a result of continuing legal difficulties over the original ownership or the rights of the peasants under the phase two regulations. A small number of landowners opted to sell their lands to the cultivating peasants under phase two. Thanks more to pressure and intimidation than to a real desire for change, 12,177 peasants chose to sell their rights in cultivation to their landlords. Several joint companies were set up by landlords and peasants. The least popular of the choices offered to the landlords was division of the land on the basis of the ownership of traditional shares in the factors of production and only 8,112 parcels of land were dealt with in this way. Approximately half a million (460,280) medium/small landowners were left in entire possession of their original holdings on the grounds that they ran mechanized units using daily paid labour or held properties too small to qualify under the regional ceilings laid down by the Land Reform Organization. Despite this marked change in phase two from radical reform of ownership to modification of tenurial conditions without necessarily a change in ownership, the reform was wide-reaching in its effects. According to the Land Reform Organization, some 1,454,187 peasant farmers enjoyed improved conditions after the reform, affecting, with farmers' families, some seven and a quarter million rural residents. In addition, peasants on endowed lands whose terms were improved under phase two of the reform numbered 79,742 on *vaqf-e 'amm*, endowments to public institutions, and 79,742 on *vaqf-e khas*, private endowments.

Somewhat before the end of phase two of the reform it became apparent that there was dissatisfaction with the progress in agricultural productivity. Looking back on the reform it was felt in some quarters, such as the Ministry of Economy, that an opportunity for the modernization of agriculture had been missed. Phases one and two in their different ways had merely reinforced the traditional sector and in many ways increased its intractability to modernization. In particular, there were fears that the new system of land ownership and farm management would lead to the fragmentation of property and cultivation rights through the effects of inheritance. To an extent the apprehension felt in Tehran was justified. The skill of the original phase one land reform law was that it had used traditional division of farmland to effect land division. Farmers were in most cases granted ownership of those plots of land which they were cultivating in the year in which the reform was implemented. The pattern of cultivation within each village ploughland or *nasaq* gave peasants, either as individuals or as cultivating teams, rights to specific areas often made up of several separate plots. From the first day of the reform, therefore, the process of parcellation of farmland was perpetuated. As time passed and plots were inherited, there was a new danger of each of the separate parcels of land being divided among heirs into small fragments, thus making its economic use extremely difficult.[27] In fact, some 14,878 villages or part villages had been bought from landowners by the year 1345 (1966/67), representing some 90 per cent of eligible villages, and rather less than that total transferred to peasant farmers.[28]

The area involved took in parts of some 30 per cent of all villages and possibly 10 per cent of all farmland.

Productivity in Iranian agriculture has lagged perceptibly behind other areas of the economy. The Central Bank suggested that growth in value added in agriculture rose on average in prices of the day at 3.4 per cent for the period 1963/64-1967/68 against an average rise in Gross Domestic Product of 8.4 per cent. In effect, agriculture was developing the least rapidly of all sectors of the economy, yet accounted for approximately a quarter of all national economic activities by value and almost two-thirds of all employment. Agriculture came to be seen as a drag on the performance of the economy at large that was polarizing the country between a modern urban sector and a backward countryside. The government, largely driven on by those with little sympathy for the farming community and with patchy knowledge of the impact of earlier land reforms, came to the decision that more upheaval was necessary, this time to consolidate the ownership and management of land into larger units that would permit rapid mechanization and an accelerated rate of growth in value added.

A first political step in this direction was made in January 1967 when a twenty-point manifesto was proposed laying the basis for a third phase of land reform. The reform was proposed to enable an end to fragmentation of land, to permit the formation of new structures on the traditional lands, to accelerate mechanization and in various ways to increase rural productivity and incomes. In 1968 articles for the Distribution and Sale of Rented Farms were enacted. In essence, these provided, first, that all ploughlands under tenancies or, later, joint farming ventures arranged under phase two were to be converted to full ownership either by sale of land to tenants or division of land between landlord and peasant. Compensation for landlords was provided at a modest level and gave no incentive for them to assist in a rapid implementation of the law. The law continued in effect until a closing date of 22 September 1971 was enforced by the authorities. The government claimed that some 800,000 tenants, approximately two-thirds of the farmers who had their tenancies improved under the second phase, received ownership under phase three regulations. The remaining tenants, for whatever reason, apparently opted not to press for the ownership put on offer.

A law for the formation of farm corporations was passed on 17 January 1968, which gave the government the legal basis on which to inaugurate its land consolidation operations under phase three. The policy linkage between the Distribution and Sale of Rented Farms Act and the Law for the Foundation of Farm Corporations was clear enough to disabuse the rural population of the belief that the grant of ownership rights in land to the cultivating peasants was an objective in its own right. The new programme transparently included both pieces of legislation, the grant of land in full ownership to the farmers being a necessary preparatory step for the creation of farm corporations out of villages occupied by these same peasant farmers. In early 1968 the Land Reform Organization was consolidated as the Ministry of Land Reform and Rural Development under its last director, Dr Valian, presumably to facilitate the setting up of farm corporations.

In theory, the restructuring of agriculture implicit in phase three permitted the establishment of joint stock companies in which the beneficiaries of land reform and, on occasions, others in the same village could lodge their land in return for an equity holding. The authorities announced in 1968 that the land and farm management consolidation process under phase three regulations would be experimental for an initial period of five years and would go hand in hand with an incentive scheme for farmers working on uneconomic fragments in villages other than those affected by conversion to farm corporation status to sell their interests.

The first farm corporation was officially inaugurated on 12 May 1968 in Fars. The Ministry made it known that it would set up three-man committees, consisting of one agent of the ministry and two elected representatives from each village to be brought into the programme. The committees were given responsibility for assessing the values of shareholders' land. An adjudicating system was also set up to arbitrate in cases of dispute. The minister in charge decreed that farm corporations could be established to take in more than one village. For the first four corporations brought into being the surface area ranged between 1,000 and 2,000 hectares. In 1968 it was envisaged that as many as 100 corporations would be established before 1973, the end of the 1968-73 five-year plan period. Special credit arrangements were made for the new ventures, which were given priority over traditional areas in access to funds made available through the Agricultural Bank. A sum of 530 million rials was allocated as an initial tranche to the Bank for this purpose. As an incentive for new capital and improved management in agriculture, a ten-year tax holiday was offered to those beginning new farm enterprises in either the livestock or arable sectors.

The initial response by the farming population to the proposals in phase three of the reform was inevitably negative. Some peasants were pleased to be offered the prospect of full ownership, which, after all, was what they had been promised in 1962 under the original Arsanjani reform. But there was also deep apprehension among both peasants and landlords concerning the government's intentions on the matter of farm corporations. Most beneficiaries of the reform could foresee that their land would be taken away again and that their rights to their cultivated land would be reduced below the level that prevailed before January 1962. The shah and the government, if not immediately then gradually but inexorably, lost all the credit with the peasantry that had accumulated through the earlier phases. The basis was thus laid for some of the damaging consequences of rural alienation from the regime which Mahdavy had forecast.[29]

Landlords of the traditional type who were not deemed to be engaged in mechanized cultivation were left to draw similar conclusions. The social effects of the third phase were, it would seem, totally unappreciated in Tehran. Most small landlords, owner-cultivators and landless labourers, the latter for the most part the *khoshneshin* or those with only manual labour to offer, comprising a very large proportion of the village population, were destined to suffer and in many cases be displaced by the reform. Taking into account the loss of many large landowners during earlier phases of the reform together with the loss of some of those who provided such services as smithying, building and trading, the third phase effectively dismantled the established social structure of the villages involved. Indeed, the transfer of labour from low productivity tasks in agriculture to provide a pool of cheap labour for the modern sector was an unstated but implicit aim of the third phase of the reform. As will be seen later, this ill-thought-out strategy was successful beyond the intentions of its initiators.

In place of traditional social organization and leadership, the government put in branches of the civil service, with the farm corporations themselves run by appointees from Tehran. Processes of centralization, already developing rapidly during the 1960s and articulated through the shah's reform programme, the spreading security system, and the growing strength of the bureaucracy in the regions, were accelerated by the adoption of farm corporation structures. The fragile but promising burgeoning of a genuine co-operative organization in rural Iran[30] was halted by the advent of the farm corporations. Official interest in the co-operative movement changed from one mainly devoted to improving the welfare of peasant farmers and stimulating self-help in the villages to one in which the co-operatives were seen as structural stepping stones on the evolutionary path to farm corporations. Declining allocation of credits through the societies, as available funds were put at the disposal of the larger commercial farms, added to the problems faced by the co-operatives and the central organization of rural co-operatives (CORC), as did competition for competent staff.

Perhaps the most damaging aspect of the farm corporation was the blighting effect it had specifically on adjacent areas but also generally throughout the countryside. The spread of uncertainty was out of all proportion to the number of corporations set up, and reached the stage where farmers were disinvesting in case they lost their assets to government-run corporations. Rural families lost faith in their future on the land and directed their children away from agriculture and the village. During the period 1968-72 the impact of rural insecurity was visible but not sufficiently bad to cause alarm. The combination of, *inter alia*, the heightening insecurity and a booming urban economy from 1973 deepened negative trends of the kind indicated to disastrous proportions.[31]

6.3 Farming structures and reform

In order to understand the special nature of the Iranian land reform and its progress following the original decree of January 1962, the 'land' element in the formula must be seen in its appropriate context within the indigenous farming structure. Cultivation was traditionally seen as comprising five factors — land, water, seed, power and labour — with varying values attributed to each depending on the region concerned and the kind of crops grown. However, no matter how much the variation in the value attributed, the status of, and the return to, the cultivator depended on the number of factors of production he provided. In some cases it happened that they were each provided by a separate individual. Thus, in consideration of the organization of Iranian agricultural production it must be borne in mind that land was only one of the major inputs involved. Many of the problems that affected the land reform stemmed in large part from the failure of the authorities to acknowledge this in the legislation laid down at the beginning of the reform programme.

In Iran, as indeed in many of the semi-arid and arid zone countries of the world, only a small percentage of the land surface area is utilized for agriculture. A mere 5 per cent or so of the country is generally under active cultivation for field and orchard crops. In the early 1960s 4 per cent (4.18 per cent in the census of 1960/61) was under the plough or down to orchards. Land as such is not, therefore, a factor in short supply. There is an apparent abundance of soil areas with fair fertility under fallow, representing in 1960/61 20.1 per cent of the total surface area of the country, or potentially cultivable (18.3 per cent). The principal problem for farmers of whatever kind is water availability either through natural precipitation or drawn from surface/sub-surface sources. Unfortunately control of irrigation water supplies was not always given adequate weight by the Land Reform Organization, which led to disputes and exploitation of land reform beneficiaries in the more arid regions of the Iranian plateau and slowed down implementation of the reform in Yazd, Kerman and parts of Fars.[32]

The pattern of land ownership before implementation of the

land reform was far from clear and there was no general agreement amongst either government agencies or independent observers concerning the relative importance of landholding groups. The position was further confused as the reform progressed since many large landowners, forewarned of the impending attack on their position by the abortive Land Reform Act of 1960, broke up their estates amongst their family or sold land to outsiders. The situation was also obscured by the Land Reform Organization, which tended to overstate the areas held by the larger landlords in order to discredit them in the eyes of the nation as 'feudal barons'.

The most useful source of information bearing on the comparative importance of the groups that owned agricultural land is the 1339 (1960/61) Census of Agriculture which showed that villages were broadly of three kinds and of more or less equal numerical (though not necessarily areal) extent — large landowner, *'omdeh maleki*, villages, medium/small, *khordeh maleki*, landowner villages, and villages of mixed ownership. Economically, therefore, the large landowners, although in a strong position for so small a group, did not dominate land holding *per se*.[33] Table 6.1 shows the estimated distribution of landownership before the decree of January 1962.

Table 6.1. Estimates of the distribution of land ownership in the period before the Land Reform Law of January 1962.

Type of ownership	Per cent of all land owned	Number of villages	Per cent of all villages
Large proprietors	56.0[a,b]	13,569[c]	34.43
(of whom those owning over 100 ha)	33.8	—	—
Small proprietors	10-12	16,522[c]	41.93
Royal domain	10-13	812[c,d]	2.06
Religious endowment	1-2[d,e]	713[c]	1.81
Tribal holdings	13.0	—	—
Public domain	3-4[f]	1,444[c,f]	3.67
Other holdings	—	6,346	16.10

Notes

[a] Hadary, 'The agrarian reform problem', suggests that 56 per cent of all land owned is in the hands of 100,000 proprietors.

b This figure is supported by Eng. Dehbud, Deputy Director of the Land Reform Organization, who quotes 55 per cent of land belonging to large proprietors.
c Official estimates by the Land Reform Organization (1963).
d Official statistics, Administration of Crown Estates (*Idāreh-ye Amlāk*).
e Dehbud *Land Ownership and Use Conditions* quotes 15 per cent for charitable endowments as a likely figure. Hadary suggests 25 per cent as the total area of *vaqf*.
f Some 1,800 villages were Public Domain in 1951. Dehbud quotes 1,754 villages in the Public Domain in 1963. The difference in these figures and the official figures noted in the table is due to the inclusion of farmstead settlements in the former calculations.

Source: McLachlan, *Land Reform*, p. 687.

A significant matter that has never been seriously examined, if actually understood, was that although the *economic* power of the large landlord class was firmly rooted in their ownership of estates of one entire village or more, their ownership of quite small areas in other villages was sufficient to spread their *political* influence far beyond their immediate estates. The political implications of land ownership, especially through this extended sphere of family influence arising from ownership of scattered and possibly agriculturally insignificant areas of land, led observers to exaggerate the land-holding roles of the large landlords. Phase one of the land reform was designed to limit the political power of the great landlord families, and in this there was probably a coincidence of views between the shah and Dr Arsanjani. Yet even here there was a great deal of confused information and policy. It was perceived that the great landlords tended to predominate in political affairs in the capital through their influence in the *majles*, the Court and the bureaucracy. From this it was deduced as a universal premise that their political power derived from the concentration of large landed estates in their hands. Yet, at the time many political leaders and holders of high office were not owners of vast areas of land or numbers of villages. Meanwhile, many large landowners played small or often negligible roles in national political life.

The situation is well illustrated by the south-eastern districts of the country during the land reform. On completion of the land acquisition and distribution in the Sistan and Baluchestan area it was noted that only two villages had been purchased by the Land Reform Organization out of a total of 1,283 villages in the area. It was assumed by Halliday, amongst others, that the local family of influence — the Alams — who had long been a powerful political force in Tehran, had blocked the implementation of the land reform in order to protect their interests. In fact, although one branch of the family held significant areas of agricultural land in southern Khorasan (Qa'enat) and Sistan (Zabol), the family influence throughout Khorasan/Sistan/Baluchestan was mainly historically based on qualities of traditional leadership, successful mediation with the central authorities and an ability to dispense patronage in the regional civil service. In southern Khorasan the basis for political influence on the part of the family was strategic land ownership rather than bulk ownership of villages. Thus, the Tabas outpost of the family empire was held through the arm of extended alliance with the Birjandi family and a small element of land ownership. In Gonabad, the family had far more influence over local events than was justified by their land and water holdings, which were almost negligible.

The case of the Alam family, indisputably among the elite of the so-called 'five hundred families' or 'thousand families', demonstrates that great care is needed in examining the linkages between land holding and political power in pre-land reform Iran. It will be clear that in the case of the Alams the land reform would not deeply hurt them since their lands were fragmented among various branches of the family and their ownership was in small shares of land and/or water in a large number of separate villages scattered over a considerable area of south-eastern Iran. Asadollah Alam retained a very considerable political role in Iran as prime minister and confidant to the shah through to his death in 1978, on which the effects of the land reform could never have touched adversely.

The author had opportunity to meet a wide range of landowners during field work in Iran during 1963 and 1964. There were few large landlords with political aspirations who had strongly antipathetic views of the land reform. They did not like its effects and many thought it badly conceived or irrelevant to the country's real agricultural needs; but none saw the reform as ending their chosen careers within the political milieu of the Court or the opposition. A majority of large landlords who were absentees either in Tehran or often abroad lost their sources of income, but by definition in most cases they had abandoned direct active interest in Iranian domestic politics long before the reform. The most angry and disaffected of the larger landlords at the time of the first phase of the reform were those with genuine farming interests but no direct involvement in the national political processes. Forceful and strident attacks on the land reform were more common both before and after phase one from the medium and smaller landowners or the religious authorities.

6.4 The organization of cultivation

The pre-reform structure of ownership was offset by an overlapping system of land utilization, which profoundly affected the owners of land coming under the terms of the Land Reform Decree of January 1962. Farmland owned by persons with more than a one-twelfth part of a village was either worked directly by the owner, put in the care of the landlord's bailiff, or rented to another cultivator. There were no figures available for the country as a whole, but on the basis of the 1960/1 sample taken by the Ministry of Agriculture and personal observation by the author, it would seem that the characteristic type of organization in Iranian agriculture was that of landowner cultivation through an agent in the employ of the landlord. In respect of the Land Reform Decree this factor was of prime importance, since share-croppers employed by landlords had more chance of receiving land than in arrangements where estates were worked by contractors. This was especially true where the contractor was the owner or traditional provider of the power, oxen, input in cultivation (*gavband*), since he rather than the share-cropper could claim land distributed in the reform.

Below the upper managerial level, agricultural organization became more complex, and varied from region to region and even from village to village. There were two important final contracts for cultivation at the village level that prevailed in most parts of the country. First, there were peasant contracts, which accounted for some 45 per cent of the farmed area and, second, share-cropping contracts, found on approximately 45-50 per cent of the cultivated land. The contracts with peasant farmers or others were mainly in the form of leases, most of them fixed for short terms.[34] Share-cropping contracts were very variable, depending on the region, the type of land or crop involved, the number of factors of production contributed by the parties to the agreement, and how just the landlord or bailiff.[35]

Among the remarkable Persian indigenous organizations for field-level cultivation was the *boneh, sahra* or *haraseh*. The *boneh* was a group of between five and seven members who worked land under contract, jointly providing the labour factor in production. The classic operation of the *boneh* has been described by Safi-Nejad,[36] while other excellent analyses by, among others, Ehlers[37] and Amini[38] are also available. *Boneh* systems were a group response in the particular social and economic conditions of certain Iranian large landowner villages, particularly in circumstances where landlords were absentees. In essence, the groups were organized by a leader, *sar boneh*, to undertake ploughing and other activities throughout the agricultural season on a crop-sharing basis with the landlord.

The effects of reform on the *boneh* were considerable. Private ownership removed the motive for joint group action in cultivation[39] and the system disappeared over considerable areas. Attempts to reinstitute it through a new land reform in 1980/81 as a means of recreating collective and egalitarian conditions in the countryside failed as a result of opposition both from the peasants and from the Council of Guardians of the Constitution (see Chapter 8).

In most areas of traditional agriculture animal power, mainly provided by oxen, was used for ploughing, threshing and other farm duties. Oxen or other means of traction might be provided by the landlord, the peasant proprietor, the share-cropping peasant or another party. Whoever made available the tractive power in cultivation usually took one-fifth of the crop in return. In a number of areas the provision was undertaken by the *gavband*. His importance in the implementation of the land reform arose from the fact that in several areas and under certain farm management conditions he was able to receive land under the terms of the law rather than the peasant providing labour.

Some observers of the reform assumed that the *gavbands* had become owners of land and that the reform consisted of nothing more than a movement of land from one set of capitalist owners to another. This view was wrong in several major respects. First, the *gavband* was not a universal phenomenon in either a vertical or horizontal sense, i.e. the *gavband* was present only in some areas and, where he was present, he operated only within certain sectors of the organization of cultivation, viz. in the landlord/share-cropper or the contractor/share cropper structures. In practice, the *gavband* was common in the more arid areas of the country, with a particular concentration noted in the Tehran Province.[40] During the land reform many *gavbands* received land but only where they were resident, had a contract for cultivation directly with the landlord, owned no land of their own, and provided rather than financed the motive power in cultivation. Those *gavbands* who gained ownership under the reform within the constraints already noted did so under the first phase of the reform which, taking the reform as a whole, affected a comparatively small part of the total area of the country that fell to the land reform in all its various phases.

The problem of the *gavband* was that where he had provided a valid function before the reform, the situation after 1963 was such that, if he had not benefitted from land distribution, he would find it difficult to continue his role. As with the *boneh*, private small-scale ownership of land after the reform removed the need for inputs from outside the family circle. Peasants could use their land to raise credits so that they could own oxen and save on use of the *gavband*, many of whom were forced to look elsewhere for an occupation.

In such circumstances, the argument of the switch of land ownership during the reform from large owners to intermediate capitalists[41] appears to be more selective and limited in its geographical spread than its supporters possibly realized. The agrarian structure in the traditional villages at the end of the reform was skewed, albeit slightly, towards peasant proprietorship and peasant small-scale tenant farming, including those *gavbands* who gained land under the reform. The costs of phases one and two were borne by the larger landlords and the landless non-beneficiaries, as the ownership and management structures were changed to favour private small-scale operations. Phase three accentuated this trend against the interests of the landless, as remaining tenant farmers gained full ownership. The main difficulty seen by the government at that stage was the parcellated nature of the ownership of farmland, often in tiny and fragmented private holdings. In many ways phase three of the land reform was used by the authorities as a preparation for the reconsolidation of farm units into farm corporations and production co-operatives during the 1970s. This would scarcely have been necessary had the traditional villages emerged from the reform with a commercial orientation or under an ownership system that enabled large or medium capitalist farm management units to be organized by the *gavbands*.

On shrine lands the arrangements for cultivation were also varied. In most villages the rights in land and water under *awqaf* were either managed by trustees for the religious authorities in the case of public endowment, *vaqf-e 'amm*, or supervised by appointed trustees acting for the individual or family in the case of private endowment, *vaqf-e khas*. With few exceptions, *vaqf* lands were in a poor state of cultivation. Official trustees held large amounts of land. Some 2 per cent of cultivated land, comprising more than 700 villages, was in public endowment at the time of the land reform. Most gifts of immovable property to public endowment were tiny fragments of land, as small as 33/1,120ths of the Fathabad Rashtkhar property in Torbat-e Haydarieh, for example, which was in trust to the Shrine of the Imam Reza at Mashhad, while tiny fractions of water shares in *qanats* were similarly endowed. The registration documents (*vaqfnameh*) of the Torbat Haydarieh and Kashmar districts to the south of Mashhad show almost every village affected by the allotment of minute parcels of land for the benefit of the shrine. Location and management of such a broad sweep of scattered fragments was a difficult problem for the trustees of the shrine to which no effective answer was offered. In consequence, shrine lands were left to management under lax conditions that brought low output from the properties concerned and a poor income for the shrine. Where larger single blocks of land were in the hands of trustees, they were auctioned off for cultivation by private contractors under fixed-term leases. A number of the operators were very efficient, though of necessity, because of the brief period of lease, exploitative of land and peasantry.

Privately managed endowment lands were no less badly run in many cases. Within a few generations of an endowment of land within a family there would generally be a proliferation in ownership so that the yield was insufficient to provide any of the beneficiaries with an incentive to maintain production. Indeed, in a number of villages in the Juimand/Gonabad group in 1963 there were parcels of endowed land whose ownership was entirely lost in the mists of multiple absentee ownership.

The land reform began tentatively in respect of endowed lands, with slight modifications to the terms of lease which improved the status of cultivators. Later public endowment was annexed to state control, and the income of the properties of some of the shrines was replaced by direct financial grants from the government. The situation of the shrine of the Imam Reza at Mashhad was different in that the shah was the *mutavalli* or guardian and revenues were largely collected by the shrine administrators. In the teeth of legal traditions, endowed lands were ultimately allocated to the cultivating peasants on the same terms as were offered to those on private estates, bringing an immediate and in some cases dramatic improvement in productivity.

6.5 Water as a factor in production

Water as a factor in agricultural production has obvious significance in a country which suffers frequently from prolonged drought, exhibits all the characteristics of the semi-arid environment, and experiences severe physical constraints on cultivation. Yet, as we have seen, only a fraction — some 20 per cent — of cropland is irrigated, accounting for approximately

4.5-5.0 million hectares, and it is on this that all the commercial crops, vegetables and even two-thirds of the wheat and barley crops are grown. Water and land ownership often go hand in hand but there are notable and widespread exceptions to this rule — in the Yazd/Kerman area, for example. Nevertheless, land and water have a co-relationship in cultivation, with certain plots of agricultural land associated with supply from particular water sources. This situation was reinforced by the system of gravity water feed from the *qanat*, stream diversion or well, which meant that land could be commanded only down-slope from the irrigation line, with specific irrigation channels serving plots in the ploughland or gardens below them within an established rotation of water distribution.

The land reform laws laid down that water supplies traditionally associated with the cultivated area of a village were to be transferred with the land to their new owners. In cases where the water was supplied by the landlord the matter was straightforward and no real immediate problems were met, other than those arising from management and maintenance of the water source. In cases where land and water ownership were separate, more difficulties were experienced, particularly where *qanats* were the main source of irrigation water. In the Kerman area, land distribution proceeded slowly while legal complexities inhibiting the transfer of land and water rights to beneficiaries of the reform were settled.

Water supplies created a long-term problem for the reform. This was particularly so where *qanats* were used, effectively involving 60 per cent of irrigated lands. *Qanats*, which rely on extensive underground water channels, are costly to construct, take many years to complete and need regular and thorough maintenance. There is a strong case for believing that the most vital function of the landlords in pre-reform Iran was the provision of and care for water supplies. In undertaking this, the landowners shouldered much of the risk in agricultural production, for the *qanat* was both a delicate and perverse instrument of water supply. Annually, the underground section was liable to damage by flooding as a result of a rainfall regime prone to downpours and storms in a land surface thin on vegetation that encouraged sheet flooding. Periodic roof-falls occurred, occasionally over sufficient lengths in individual *qanats* to close them long enough to cause crop failure in their service areas. Earthquakes aggravated this situation. The earthquake of 1963 ruined or partially blocked *qanats* over wide areas of the Qazvin Plain. Long-term fluctuations in the water table and even concurrent years of drought led to a decline in water supplies in *qanats*. Beginning before the reform a number of landowners invested in deep wells, lifting water using diesel sets. The water table dropped dramatically in some areas — Ferdaws in Khorasan, for example — in such circumstances, driving the water table below the catchment zone of the well system serving the established *qanats*, which dried up. In all, the provision of water in pre-reform Iran could be a costly and hazardous business for which the landlords bore a heavy but generally unregarded responsibility.

The land reform regulations provided that the running of central water supplies such as the *qanat* would be taken over by the cultivating peasants working through a new co-operative society. At the time it was clear that this assumption was questionable. Most villages were slow to make their co-operative organizations work and even where there was goodwill adequate for the co-operative to act efficiently it was not always possible for the peasants to carry the magnitude of risks involved in repair and maintenance of their *qanat*. Official agencies controlling credit were in many cases reluctant or had inadequate financial resources to lend money to village co-operatives for so expensive and uncertain a venture as the repair of a *qanat* badly damaged by floods, particularly, as was likely, when a dozen other villages in the same district were similarly afflicted. The confusion over the 'water' component in the land reform took many years to resolve, and it was not until the National Water Authority began serious operations in the 1970s that a pattern of control and management emerged in the traditional irrigated areas, by which time a considerable amount of damage had been done. The role of the *qanat* had been considerably diminished by modern irrigation project developments.[42] This was a process that had the tacit support of parts of the bureaucracy.[43] There seems little doubt, too, that the poor provision for irrigation water supplies in the wake of the removal of the large landlords, and the loss of confidence to invest in this area by those landowners remaining in agriculture, did much to reduce the strength of farming in traditional villages.

The shah's speech from the throne in 1967 contained an announcement that nationalization of water resources would be dealt with in the near future. It was expected at that time that the state, which already owned the large rivers, would take ownership of all water supplies, including rivers, *qanats*, wells, sub-surface reservoirs, lakes and streams. The government's short-term aim was to assume powers that would enable it to impose its policies for water use, though eventually to create local control of individual water systems through government administration of them. Above all, the authorities hoped to end the division between private owners of water supplies and the farmers on land using such water, which had so bedevilled the land reform in its earlier phases. There was the hope too that available supplies could be utilized more efficiently by a combination of regulatory measures and fees levied on water use. The growing costs of state water storage schemes and their associated irrigation works encouraged the government to feel that unrestricted water use in agriculture was profligate of a limited and expensive resource.

6.6 Agricultural credit

In the first year of the reform credits made available to farmers on distributed land were very limited in amount and were valued at a mere 6,800 million rials or 19 per cent of the budget forecast for agriculture during the entire period of the Third Plan 1962-8. At that time it was assumed that landlords either had never performed a major role as providers of credit or that they and the traditional sector, including the bazaar, would continue to provide credit in spite of the reform programme. In its early stages implementation of the land reform made slow progress and only a fraction of the peasants were actually affected. In 1963/64 the provision of credit facilities was first undertaken by the Agricultural Co-operative Bank, but on a modest scale since the bank was given insufficient credit by the government. Only in 1964/5 were there moves to set up an organization within the bank that could cater for the allocation of credits on a large scale. To an extent, the hand of the authorities was forced by the fact that a large number of peasant families had by this time received land under the reform and were facing a credit crisis. Weather conditions, including a disastrously hard winter that penetrated even the central deserts and a drought-blighted spring, had been severely adverse to cultivation and had put the whole reform in jeopardy. In the first six months of the Iranian year 1964/5 the Agricultural Co-operative Bank made 900 million rials available in credits, and 4,500 million rials in the second. Such a response might appear generous but, on average, only 10,000 rials (approximately $153 in money of the day) was paid out per successful applicant.

Erratic and patchy as the credit situation was in the early years of the reform, there was at least dedication on the part of the banks to the interests of the small farmers. The declaration of the third phase of the reform in 1967 changed this situation quite radically. Government policy became increasingly oriented towards promotion of a commercial agricultural sector. The members of co-operatives had to make do with mainly small-scale short-term credits through the Agricultural Co-operative Bank. Long-term investments by individuals were funded through credits drawn on the Agricultural Development Fund, later changed to the Agricultural Development Bank, set up in 1968 specifically to assist commercial farming, or the commercial banks. Credits from these latter two sources were mainly steered towards the larger farms, especially the newly established agro-industrial units or farm corporations.

Traditional farmers were increasingly badly served by the official credit system and were left to fend for themselves in the informal sector, which had always provided the bulk of credit for agriculture. By 1970 total credits demanded by agriculture were estimated at 60 to 65 billion rials, of which the formal system contributed some 20 billion and the informal sector the balance.[44] The costs to the small farmers who were the principal users of the informal credit market were financially high, though there were compensating factors such as marketing advice available through the bazaar merchants.[45] In the Iranian year 1973/74 some 2.3 million members of co-operatives shared a total credit from the Agricultural Co-operative Bank of approximately 20,000 million rials. In the same year 310 commercial farms drew 2,650 million rials from the Agricultural Development Bank. By 1978, by which time agro-industrial units were in operation, the swing had become more acute, with 47 per cent of credits to the traditional sector via the Agricultural Co-operative Bank and 35 per cent to the large farmers through the Agricultural Development Bank. Taking into account loans from the commercial banks and investment and equity taken by the state or state-funded agencies in private companies and institutions, a small number of modern farm units took more credit absolutely than the entire traditional sector.[46]

6.7 Reinforcement of commercial farming: agro-industry

In the search to find new means of improving agricultural productivity the government was encouraged to undertake rapid development of the lands lying in the service area of the major water storage dams.[47] This was made possible by legislation in 1968 which enabled the Ministry of Power to allocate land for agricultural development downstream of dams it had built. There was a belief at the time that agro-industry was an ideal vehicle for modernization using an integrated organization of large-scale farming and associated crop-processing industry.

The concept was first envisaged for Khuzestan where water supplies were available from the Dez dam built during the Second and Third Development Plans with the financial assistance of the International Bank for Reconstruction and Development at a cost estimated at $95 million, a very large sum at that time for Iran. The generators at the dam supplied 2.20 Mw/hours of electric power for the electrification programme in the south and for the national grid. The authorities had been very slow, however, to make use of the irrigation potential of the 3,467 million cubic metres of water available from the reservoir, especially since the project was originally conceived as a multi-purpose project and would only become fully economic when irrigation revenues were earned.

From the beginning every effort was made to bring in foreign interests to participate in the scheme. The Dez irrigation project was advertised by the Development and Resources Corporation in the USA and elsewhere as one where the 'unusually advantageous combination of land, water and climate makes the agricultural potential of Khuzestan comparable to that of the Imperial Valley of California... the Project can become... one of the most highly productive agricultural areas in the world'.[48] The involvement of foreign venture capital and the mobilization of expertise had much to recommend it, though it might be argued that too much was paid in credits by the Iranians to induce foreign companies to enter the project and that too much return as measured by production was expected from the agribusinesses by the Iranian authorities. Expectations on both sides were to be severely jolted before the close of the project.

The Dez dam was completed in 1963 and a pilot agricultural scheme with a gross total of 21,500 hectares was developed on lands below the dam. It was not until 1967 that the second phase of the scheme was begun, to develop an irrigation network servicing 74,792 gross hectares. Pah-e Pol block was completed in 1969, after which the rest of the area was converted culminating in the Dahli and Shahvali areas in 1973. In all it was expected that 68,000 hectares would be available for development under agro-industrial units. By 1971 there were five units holding 30-year leases on some 66,000 hectares, including H.N. Iran America Agro-industrial Company, Iran California Company, Iran Shell Cott Company, Dez Kar Company and the Ahwaz Sugar Refining Company. Additionally, an Iranian organization, the Shahvur Company, leased 5,000 hectares under the Khayrabad dam in Khuzestan.

The successful recruitment of private ventures to Khuzestan opened up the prospect for similar developments elsewhere. Water from the Aras dam, jointly constructed with the USSR in Azarbayjan, was available for irrigation on up to 60,000 hectares in Dasht-e Moghan. In the Jiroft region some 4,000 hectares were designated for indigenous agribusiness development and other areas under storage dams were also examined for the same purpose. The final solution — agro-industry in which peasants had no role whatsoever — was offered under the Fourth Development Plan by the Ministry of Water and Power. Indeed, during the Fifth Plan period, 1973-8, the government adopted a parallel policy for the traditional areas under which the country's investment resources in rural areas were to be allocated exclusively to selected 'growth poles'.[49] The plan was to concentrate 12,000 villages into 1,180 rural development zones centred on selected growth poles in which the new farming structures and all their support services would be located. Existing co-operative societies on 600,000 hectares in the new growth poles were to be made into production co-operatives before transformation to farm corporations. Modern housing, electricity supplies, schools and a host of other facilities were to be set up in these centres with a view ultimately to creating as many as 7,000 to 8,000 poles. All villages and settlements with populations of less than 250 persons were to be excluded from participation in social and economic development projects sponsored by the state. Application of the new criteria would have meant 48,000 villages and farmsteads, involving 26 per cent of the rural population, being abandoned to their own devices. Only the government's slide into severe economic problems in 1976 and a virtual moratorium on non-essential development schemes in subsequent years saved the countryside from the steady application of this growth-pole programme.

Government preoccupation with both agribusiness and farm corporations was rarely justified. They comprised small proportions of the total agricultural land. Both carried severe penalties — agribusiness through the displacement of village communities formerly occupying the land they reclaimed[50] and farm corporations through their blighting effect on traditional farming. The misplaced effort of the official institutions is even more accentuated when account is taken of the fact that the combined contribution of the two modern farm structures to total output[51] in 1975 was only 2.4 per cent against 95.9 per cent from traditional farms and 1.7 per cent from irrigated commercial farms. These estimates are presented in Figure 6.1. Extremely scarce resources of time, funding and management

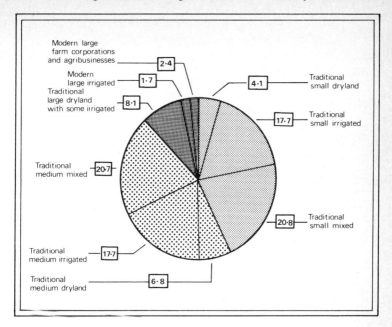

Figure 6.1. The contributions of traditional and modern sectors to agricultural output 1975.
Source: Booker Agricultural Services, *National Cropping Plan*.

were devoted to farm corporations and agribusinesses which might have been more effectively deployed elsewhere in agriculture. As the flow of credits to the agribusinesses was reduced in the mid-1970s their position deteriorated rapidly, a trend exacerbated by acute environmental constraints, poor management and antipathy to them on the part of the local people. By 1978 all but one of the agribusiness units had accumulated unsustainable losses and their thirty-year leases were virtually abandoned.

6.8 The regional effects of reform

Implementation of the land reform reflected the rich regional diversity of Iran. The regulations for phase two provided a range of ceilings on the area to be owned by landlords buying out the rights of peasants (see Figure 6.2).

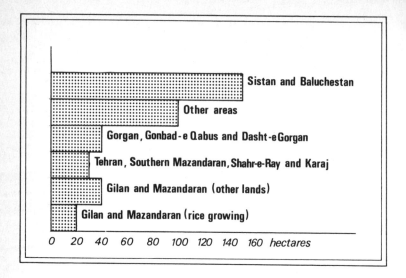

Figure 6.2. Regional variations in maximum size of holdings in land purchased from share-croppers under phase two regulations.
Source: D.R. Denman, *The King's Vista*, p. 136.

Under phase one, activities were regionally skewed. For example, 509 full villages were acquired in East Azarbayjan against a mere one village in Sistan and Baluchestan and none in Yazd. There is no simple way of ranking the provinces by the area affected by reform since all lands were measured in traditional systems that are not tied to absolute values. Figure 6.3 indicates the number of beneficiaries under phase one by their province of registration and an index based on numbers of properties/price of land (average price per unit of land purchased by the Land Reform Organization).

Using this scale it can be seen that the reform had its greatest effects in the north and north-west of the country and was least felt in the centre, south-east and Gulf regions. The comparatively low rating of Fars arose from the problems of implementation in this essentially tribal province, where 221 part and whole villages remained in dispute even as late as 1972. Despite the variations in the incidence of phase one, it is clear that in all but a few poor and isolated areas of the east and

Change and Development in the Countryside

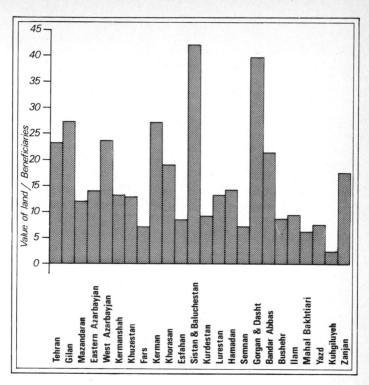

Figure 6.3. Regional effects of phase one of land reform to October 1972.
Source: Land Reform Organization, Tehran, 1973.

south, the demonstration effects of peasants receiving full rights in ownership of land were widespread. Throughout the country peasants who had not been involved in phase one were unsettled by seeing their fellows actually receiving allocations of land. Rents were withheld from landlords in many districts by peasants attempting to force application of the reform law in their own villages, and on occasions violence was used. In practice, therefore, once phase one was seen by share-croppers to have been put into effect, it would have been difficult for the government to stop the reform continuing.

The incidence of phase two was all but universal, since it involved most medium and small landowners, and they were represented across the country. No attribution of values by way of surface area or market prices of land involved in phase two was undertaken. The Land Reform Organization issued only totals of land transactions completed categorized by the type of option taken by landlords — lease, sale, purchase of peasant rights, joint enterprises or division (see Figure 6.4).

The effects of phase two were largely predictable. It was the very universality of phase two and the subsequent phase three that was the key to the geography of land reform between 1963 and the end of the 1970s. Very few villages escaped the net of government action in phases two and three, unless they were designated as mechanized and/or did not employ share-cropping labour or the landlords were able to use financial and political influence to protect their interests.

In the fertile areas of the country beneficiaries of the reform were able in the years preceding the announcement of phase three to establish themselves in viable units. Undoubtably there were problems even in these richer villages. Indebtedness to the merchants and the itinerant salesmen (*pilevars*) was inevitable since the co-operatives were unable to give adequate financial support to their members.[52] Uncertain weather was a constant difficulty. But for the most part the newly emancipated farmers in the fertile districts such as Gilan did well from the reform. In the poor rainfall areas the situation was rather different. Removal of the guiding hand of the landlords and their ability to raise funds for agricultural projects, a low level of ability to pay off loans from low-yielding activities, and greater exposure to natural hazards made their positions extremely tenuous. In such areas the poor rate of formation of co-operative societies, and the inefficiencies of those that were set up because of shortages of funds, officials and peasant support, were at their most damaging. The geographical divisions of the country took on new meaning as the later phases of the reform were implemented, with differences between the richer and the less well endowed areas increasingly accentuated. In the latter regions, many families, whether beneficiaries under the reform or *khoshneshin*, found little incentive to stay in their traditional occupations, especially as the urban areas grew prosperous in the period after 1972. In practice,

Change and Development in the Countryside 141

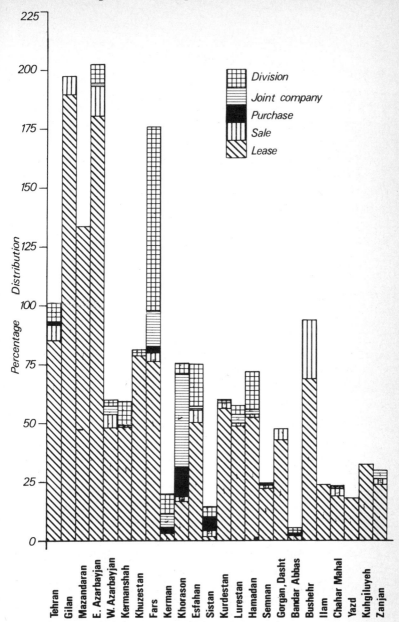

Figure 6.4. Regional distribution of transactions in phase two.
Source: D.R. Denman, *The King's Vista*, p. 338.

there emerged a polarization in agriculture as a result of the reform which severely loosened the grip of the rural communities on those village-cultivated lands in environmentally disadvantaged districts — essentially those of the centre, south and east. The onset of rapid urban economic change in the 1970s therefore wreaked much more thoroughgoing effects on such areas than would have been the case had reform of the kind adopted by the government not taken place.

In retrospect it is clear that phases two and three in particular made the countryside extremely vulnerable to later events. While the authorities could not have foreseen that this would be so in the mid-1960s, it was certainly apparent in the early 1970s and could have been taken into account before the feckless spending spree was set in train in the 1973-78 plan period that proved so damaging to agriculture.

6.9 The shah's reform programme and changes in rural society

Economic and demographic pressures for social change in the countryside were greatly increased following the reforms of the early 1960s. It may be argued that there were two critical periods involved, first, the years 1962-72 when changes were induced by the land reform and its associated upheavals and, second, 1973-8 when the forces working for change came from the country's exposure to rapidly augmenting oil income. Inevitably there were elements that overlapped the two periods: the pressures of strong economic growth permeated the rural areas before 1973, for example. Most rural communities were engaged in absorbing the impact of reforms in the first decade through processes that were slow to reach a conclusion and which for the most part were comprehensible because they were expressed in agricultural or village social terms. After 1972 the developments were almost all exogenous, economically irresistible and rarely understood by the groups which they affected. Throughout both periods, any internal rural dynamic that might have existed was overwhelmed by the scale of upheaval forced upon the villagers from outside.

Social life was considerably affected by the set of reforms that were instituted or co-opted by the shah, beginning with the land reform in 1962, though properly having their roots in the crown lands distribution programme that preceded it. Five additional measures were combined with the land reform and put to a general referendum in 1963. The six items, initially called the Sixth of Bahman reforms, were later made the basis of the shah's 'white revolution'. Other items were brought into the programme in 1964 and 1967 so that *inter alia* it eventually comprised:

1963

i) Redistribution of land to the cultivating peasants;
ii) Nationalization of forests and pastures;
iii) Sale of shares in state industries;
iv) Industrial profit-sharing for workers;
v) Enfranchisement of women;
vi) Establishment of a Literacy Corps;

1964

vii) Formation of a Health Corps;
viii) Formation of a Rural Extension and Development Corps;
ix) Establishment of Equity Courts;

1967

x) Nationalization of water resources;
xi) Urban and rural reconstruction;
xii) Administrative and educational reform;

The reforms were important above all else because seven out of the twelve of them were immediately directed at conditions in the countryside. Equally significant, most of the reforms, though particularly the early elements within the programme, were put into effect in good faith and affected all but a few country areas. In this way, the programme contrasted with earlier measures of change that had failed at the implementation stage. The reforms brought very specific social changes in the village communities mainly through the operation of the land reform laws throughout their various phases. Less dramatic but none the less significant alterations in the social climate were set in train by the exposure of villagers *en masse* to new educational and health services and the ideas of the people who came to promote them.

The land reform deeply affected the social structure in the rural areas. Landlords of the traditional kind were removed piecemeal from their positions. The so-called five hundred major land-owning families were hit hard by the first phase of the reform and were eliminated gradually thereafter. The medium and some so-called small landowners together with the bailiffs and administrators of *vaqf-e khas* were eventually dispossessed in phases two and three. Established leadership was all but eradicated by the early 1970s. At the lower end of the social scale, the *khoshneshin* or landless who had no rights to use the village ploughlands (*haqq-e nasaq*) lost their economic role as general labourers as activities were concentrated in the hands of family groups who, unlike the landlords, had neither need nor responsibility for the landless labourers. The *khoshneshin* emigrated to the towns in large numbers with the coming of phases two and three. There were further absolute losses from the economic, and finally from the social, fabric of the villages in the form of those who were forced to drop out of the cultivating system through an inability to manage changed circumstances. Amini, for example, cites the case of Zenjiabad in Fars, where some smaller owners sold their *haqq-e abih* or rights in water since they could not raise the capital to keep abreast of the need for new investment after the reform.[53]

During the land reforms the *boneh*, which made up the most basic of socio-economic arrangements in the village, was seriously undermined and for the most part disappeared. Private ownership of land by cultivating peasant proprietors took the place of the joint teams and the complex hierarchies they had sustained. In many villages these effects came in the mid-1960s, though in the case of Zenjiabad, an endowed village, they were not felt until 1972.

For many Iranians the disappearance of the *boneh* and its various regional equivalents represented a severe undermining of traditional communal social values. In the broader context it could be added that the internal social and economic changes wrought by the land reform within the beneficiary group merely reflected the generally marked revolution in the villages brought on by the losses sustained by the non-beneficiaries. Whether designated as 'feudal', 'rent-capitalist' or 'traditional', the pre-reform social structure was swept away by the land reforms throughout much of the country. The thoroughness of the assault on the established systems derived from a constantly changing mix of factors that varied from political conviction about the reform on the part of the Land Reform Organization in its early days to the inability of successive governments to understand the need for long-term stability in the agricultural cycle. The prolonged involvement of the state in producing new legislation, institutions and policies for the agricultural reform in a period that extended from 1960 for almost twenty years helped to ensure that few villages escaped the sweep of perpetual change. The developments that were occurring concurrently in other sectors of the economy were influential in both speeding the processes of change in the villages and exerting extra pressures of their own on an already unstable situation.

In areas of pastoral nomadism the land reform had social effects dissimilar to those observed in the sedentary villages. It affected the private holdings of the tribal chiefs as in the cases of the Bakhtiari and Qashqa'i, and worked towards reducing their incomes. Tapper[54] considered that the land reform undermined the final hold of the tribal chiefs in the Shahsevan territories of Azarbayjan, while Lambton[55] similarly observed it as a further and possibly more threatening intrusion of the central power into the domain of the khans. The combined effects of loss of traditional authority, fear for their economic position, and a desire to protect the tribal way of life led to a strong opposition to the reform, not least in the lands of the Qashqa'i in 1963. Failure of this stance against the regime in Tehran merely strengthened the resolve of the shah and his ministers to break up the tribal pastoral society once and for all. The events in the villages in the wake of the reform programme were in large measure also experienced in the tribal territories, excepting only those at great distances from the regional centres of administration.

The impact of the shah's reform programme, other than the land reform, was insidious. Improvements in education were a notable feature of this process. Until the 1960s schools were concentrated in urban areas and little had been done other than at the scattered village koranic schools to reach the vast rural population of the country. The literacy rate was estimated at 16 per cent in 1960,[56] which compared unfavourably with Turkey (38 per cent) on which Iran had modelled itself earlier. By the time of the revolution in 1979 the rate had improved to some 50 per cent, though still urban and male-biased.

The main instrument in curing rural illiteracy was the Literacy Corps made up of young people of national service age who were drafted to the remote villages to carry education to their inhabitants. The impact of the Literacy Corps, which began work in 1963/64, was widespread and carried significant social implications. Villagers learned from the young men who came to their communities how to manage their affairs vis-à-vis the claims of the former landlords and the bureaucracy. They began to develop new forms of self-help in building schools, baths and other civil amenities. The spread of literacy was mainly among the young, which enabled them to be more critical of the village than they might have otherwise been, and was a factor, amongst others such as recruitment to compulsory national service in the armed forces, in stimulating the exodus from the villages to better pay and conditions of work in urban areas.

As a result of the new points of the reform programme, introduced in 1964 the work of the Literacy Corps was augmented by two other organizations — the Health Corps and the Extension and Development Corps. They were responsible, respectively, for taking flexible and low-cost services to the villages in health and agricultural development. The activities of the Health Corps were inhibited by the small numbers of trained doctors and ancillary staff available, and operations were often at the domestic hygiene rather than the medical level. The Extension and Development Corps, in particular, faced an uphill task. Most of its personnel were from urban areas and had no real understanding of farmers' problems. They found themselves isolated and generally unwanted. Individual members with skills in agronomy, veterinary care, accountancy, or maintenance of machinery, on the other hand, were well received. Some in this category played critically important roles in helping the local co-operatives, acting in effect as their secretaries and often keeping up the links long after their service was completed.

Land reform was a controversial issue from its inception. The principal actors of the early period were seeking political objectives — the removal of the landlord class — though in the cases of the shah and Dr Arsanjani for different reasons. Inevitably, however, the reform could only remain a political success if its economic basis was sound. Unfortunately the criteria for economic viability were constantly changed by the government from those applicable to small peasant farms to those demanding large-scale management units. Worse for the beneficiaries of reform, the official organizations rarely appeared to understand that the restructuring of agriculture would have a lasting effect only if the individual farm units created under the reform were economically sound. In consequence, the allocation of credits, the back-up for the co-operatives, and the provision of other basic support services were never made available in the volumes necessary to assist the new farming class. As in the case of Zenjiabad,[57] far too many beneficiaries of the reform either sold out to their family or neighbours or, according to Ehlers, fell into debt and lost their lands.[58] A severe difficulty with the lack of a clear new command structure in the villages was the care and maintenance of communal facilities such as *qanats*, irrigation canals, wells and drainage lines, which had previously been managed by the landlords and their agents.

Despite these problems, which were much less widespread than is often suggested, for the most part peasants who were given land with adequate water supplies made a success of their holdings. The sum paid each year to the Agricultural Co-operative Bank was less than the value of crops previously taken by the landlord. The author observed that many new owners worked far harder after the reform than previously, which brought increasing output. Indeed, the levels of farm production showed no fall during the implementation of phases one and two, though the cropping pattern changed a little as some land moved into crops designed for on-farm consumption at the expense of cereals. In general, it was believed in the years before 1967 that the peasantry had done well from the reform.[59]

The authorities' economic ineptitude in the management of parts of the reform was far less damaging than their blindness to the social consequences inherent in the changing pattern of land ownership effected by the land reform legislation. There was rejoicing at the obvious gains made through the emancipation of the peasant from attendance on the landlord, but there was also neglect of the welfare of the many other inhabitants of the village who did not gain land but who stood to lose from changes in land ownership. There was a great loosening of economic and social ties within the village: mutual dependence was reduced; the common enemy (the landlord) had gone; the outside world was brought a lot closer in the form of the bureaucracy, the Literacy, Health and Extension Corps and the need of individual farmers to deal directly with the offices of the ministries, the Agricultural Co-operative Bank and the urban merchants. For as long as the reform was contained within modest proportions with specific local objectives, it appeared to work without creating insuperable social dislocation. As its activities spread under phase three and the villages were exposed to growing external pressures, so the social fabric began to break up in many rural settlements. This was particularly notable as the joint impact of phase three operations and the expansion of the oil economy was brought to bear after 1972.

It might be concluded that reform was necessary in rural Iran; commentators from all points in the political spectrum agreed that this was so. To be of any significance it had to change at least one fundamental component in the agrarian structure, and land was the obvious one to choose. Yet, in changing land ownership it was inevitable that all other elements would change their relative positions in the social as much as in the economic hierarchies. It might be suspected that there had been no awareness in official quarters of the social effects that would accrue from land reform — where the social change might lead or what social and political costs would be involved. Given that the reform was mainly politically inspired, the absence of forethought in that arena, other than possibly on the part of Dr Arsanjani who expected the village co-operatives to make good these losses if only in the longer term, was remarkable.

6.10 Conclusions

A number of conclusions arise from Iran's experience with agrarian reform and development in the years 1960-78. The premises on which the changes were based were proved to be mistaken. Altering one variable within the rural economy was inadequate to bring about a sustainable political realignment to the benefit of the shah's regime, without inspired and imaginative management. Once Dr Arsanjani had gone, that latter resource was clearly missing. Watering down the original reform in phase two was a tactical mistake since the landlord issue could not be segregated between large and small landowners in the minds of the share-cropping peasantry. Land ownership for one group of peasants but continuation of landlord rule in the next village was an unhappy combination. Assumptions that co-operative organizations could be set up to give an instant form of structure for articulating command, leadership and economic underpinning to the villages in place of the landlords were naïve and disastrous in many cases. Allocating land in full title to the peasants in phase three was a transparent act of deception that was seen through by all but a handful of farmers. The linkage between the provisions of phase three and the way they enabled the government to consolidate peasant lands into large farm units was widely understood and deeply disliked.

The most damaging assumption was one that underlay all the others — that Iranian farmland was a resource like any other and existed in isolation from those who irrigated and cultivated it. The argument against the official position must be seen in two parts. First, the Iranian village system was not divisible, even though there might be separate and rival co-systems within one village. Economic and social structures were deeply interrelated and depended on a set of hierarchies which brought with them benefits and responsibilities for the management of land, labour and water for arable or grazing on those parts of the village for which the joint group had customary rights of ownership, usufruct or charity. To take away one component part of the system inevitably induced changes throughout the whole. Operation of the land reforms and the subsequent programmes for consolidation into large farm units began by removing the landlords but eventually affected adversely, and in many cases mortally, all other layers in the social and economic hierarchies of the villages they touched. Under the terms of the Fifth Plan all but two million hectares of farmland would have witnessed the *de facto* elimination of the remaining element in the village — the cultivating peasant farmer for whose benefit the reform was originally conceived.

Second, the government took the view quite early on in the reform and officially from 1967 that there was no imperative organic linkage between the village as an economic entity and the land it cultivated. It was believed that managers who had been trained externally, in the sense of extra-regional or non-village as well as Iranian personnel educated in foreign institutes of agriculture, could replace traditional farming administration in order to introduce modern practices of mechanization and the use of fertilizer and pesticides together with improved rotations and higher-yielding crop varieties.

As early as 1970 the author visited farm corporations in the west of the country where the share-holding peasants in their farm corporation were effectively excluded altogether from its operations, cheap labour being trucked in from adjacent towns and villages. Substitution of managerial personnel sent to the farm corporations by the ministry for traditional farmers in smallholdings with local knowledge of cultivation grew out of frustrations in Tehran at the slow pace of agricultural growth and the feeling that use of modern techniques could cure the problem. The arrogance of the farm corporations solution to low growth rates was typified by the way in which sites were selected on a 51 per cent majority vote by farmers in designated regions, even though individual villages were firmly against inclusion of their land. Many villagers felt, with a great deal of justification, that their recently acquired lands had been stolen from them by the government. The alienation of the peasants by what was, in effect, enforced consolidation of small farms became a considerable obstacle to mobilization of the agricultural workforce in support of more rapid improvements in production, bearing in mind that more than 85 per cent of output came from the small and medium-scale traditional farms and a mere 2-3 per cent was generated by farm corporations.

It is possible that the official approach to agriculture in a country with a sophisticated infrastructure of industrial technology, research and development and management capacity would have worked, albeit at a high cost. Iran had none of these advantages. Worse, the environment was difficult. The 'Iranian' climatic area as recognized by geographers and climatologists is unique and characterized by extremes in precipitation and thermal conditions associated with marked changes in regime with altitudinal and horizontal movement. The complex agricultural systems built up in the Iranian villages enabled survival in such a hostile environment and were an adaptation to the insecurity imposed by the country's turbulent history.

The landlords in many cases acted as agents for the government, as the ultimate resort of financial risk and as heads of the command structures within the villages. A complicated arrangement of division of activities in cultivation into five elements — land, water, seed, labour and draught animals — operated in the Persian rural cultural system. It has obvious virtues where risks are high, where physical effort in cultivation is considerable and where skills of a high order are needed in every stage of farming. Specialization, in effect, permitted survival. The existence of the *boneh* and the reasons for its central role in traditional agriculture revolved around the need for intensive skilled teamwork in labour inputs to irrigation and tillage, given the form in which irrigation water originates and is managed in final use and the heavy labour demanded in cultivation.[60] Even the *khoshneshin*, the landless, were vital components of the system, providing special services such as those of the blacksmith or the occasional peak season labourer.

The strands of the web that held rural structures and skills available to enable the management of local difficulties in a hostile agricultural landscape made the land and the community into an inseparable unit of production. Substitutions were successful but only where traditional skills were brought in to augment or replace other varieties of traditional practice. Historically, many individual landlords had been removed, affecting considerable areas, and many migrant workers had adapted to new areas as in the case of the Baluchis and Sistanis in the cotton-growing areas of the Caspian coastal plain in the 1960s and 1970s. But changes of these kinds were generally superficial: they altered little of the existing structures. It became tragically evident on the other hand that modern management and high technology production in the farm corporations and agribusiness farms were dramatically unable to overcome the problems that the traditional system had learned to come to terms with. This despite the proven records of success of these same agribusiness companies elsewhere in the world. The evidence, therefore, indicates very strongly that the government's increasingly unheeding and oversimplified approach to rural development, together with the assumptions on which it was based, were in grave error. Neglect of agriculture through indifference in the period before 1960 was replaced by negligence through misguided government policies for the subsequent twenty years. The damage incurred during the latter period was the more severe because official ineptitude in rural policies was inadvertently but deeply aggravated by the government's mismanagement of its rapidly growing oil wealth from the early 1970s.

7 Agriculture, Oil and Development Planning

7.1 A review of planning

Development planning in the modern sense was adopted in the aftermath of the Second World War when the first essays were made in setting up formal investment programmes over defined periods of time. Such planning was both an economic and a political tool. There was a genuine and widely felt belief that pulling the country out of its poverty and backwardness could be achieved with benefits for much of the population. But modernization also had political benefits for the shah since it gave him a positive programme around which he could gather augmenting support, a lever with which to exert pressures against his opponents and, if successful, a means of expanding his influence abroad.

There were many in the country who had no interest in economic change. The religious classes had lost many of their economic privileges as a result of Reza Shah's reforms[1] and were aware that they would suffer further reductions in their traditional sources of finance as agriculture, commerce and industry were removed from their traditional areas of control, i.e. from village farming and the urban bazaars. The bazaar merchants, though not universally antipathetic to modernization of the economy, were governed by traditional loyalty chains of family and religion that tended to make many opposed to what was seen as the shah's promotion of competing interests. Most bazaar traders dealt in goods such as carpets, fruits, foodstuffs and other traditional commodities that were relegated in importance by the new emphasis on modern industrial products. At the same time they were generally aligned against the shah on political grounds.[2] The landowners were by no means so uniform a group as the religious classes and the bazaar merchants, but certain of them did compose a strongly antagonistic alignment against modernization in so far as it could affect ownership of land and the status that accrued from it. All three opponents of economic change were involved in various degrees in the unrest of 1963.

From the outset, the fact that economic planning was equated with the modernization processes induced by the shah for political reasons caused a large and influential section of the population to oppose the plans. The association in many Iranian minds between the economic plans and westernization of the country was unfortunate but powerful. Americans laid down the framework of the First Plan[3] and the United States and its advisors, or Iranians trained in its institutions of advanced education, appeared constantly thereafter to be in the van of planning activity. Nowhere was this more true than in the planning of agriculture, given the roles of the US aid agencies, the political support for land reform from the US diplomatic service, and the introduction of US agrarian structures such as agribusiness.

Although a great deal has been said and written about agricultural planning, the irony is that most of the important programmes for the sector, especially in the period 1960-73, were conceived and implemented outside the economic plans themselves. The land reform, for example, was instituted entirely without reference to the current plan.[4] Not until the Fifth Plan, 1973-8, were the full range of government policies for agriculture taken into account by the planning authorities, by which time the formal planning processes were overwhelmed as government expenditures rose rapidly on schemes that were totally outside the development plan and at a pace that was known by the planners to be unsustainable. Agriculture, in consequence, suffered more than other sectors from the abandonment of the formalities of planning during the fifth plan period.

The Plan Organization too often saw the problems of agriculture in simple resource terms. Financial allocations went towards specific irrigation or production projects, while the overall requirements for general credit provision, extension activities and security of tenure were not provided for. There were many reasons for the planners' inability to elaborate an integrated and truly national policy. Responsibilities were split so that the Plan Organization had a limited voice in the counsels that governed developments in the countryside. The information base was persistently weak before the completion of the various stages of the National Cropping Plan of 1975[5] and even then far from complete in respect of management systems, skill availabilities and village social affairs. In view of the regional variations in climate, water availability and farming types, the poor understanding of rural Iran by those officials laying down the plans for its development was a severe disadvantage.

Among the effects of the technocratic approach to agricultural investment and the natural bent of planners to use their resources to find ways of meeting quantifiable production targets, were an emphasis on defined regional projects for water resource development and the establishment of development schemes on new sites as an alternative to traditional farming. Such identifiable investments as the great dam reservoirs of the Second and Third Plans were not without their merits, but rarely appeared to bring benefits to the mass of farmers in traditional village agriculture. The dominance of the production-centred approach to development eventually culminated in the regional poles policies of the Fifth Plan with all its insensitivities to both existing farming systems and the constraints posed by the environment.

The acid test of the government's ability to plan development in the post-1953 period was that it should harness oil revenues to net productive investments, which would increase the overall capacity of the non-oil sectors to provide employment, augment output of goods and services, provide sources of exports and offer the prospect of a viable economy regardless of the fate of the oil sector. The aim of applying oil revenues to development coincided with official objectives as expressed in the preambles to the various development plans.[6] Agriculture had an obvious role within the programmes for economic diversification, food self-sufficiency and the maintenance of employment and production chains. The Iranian authorities can be found wanting in the attainment of all aspects of their self-imposed economic targets. But in agriculture the shortfalls were apparent much earlier than in the rest of the economy and proved to be of disastrous proportions by the end of the fifth plan in 1978.

As we shall see, the country increasingly became an oil-based economy. Failure to manage the impact of oil revenues was felt most in urban areas with effects such as high rates of rural migration. Official preoccupation with projects in industry, defence, and prestigious public building projects meant that the attention and interest of the shah and his ministers were drawn to horizons increasingly distant from peasant agriculture. Oil provided both government and population with an easy low-effort alternative to grappling with the chronic difficulties of traditional agriculture. It left the demise of the village farming community as an opportunity cost of 'high growth' and urban-based economic expansion established on the uncertain foundations of a high oil income.

Political dedication to agriculture became enfeebled once its benefits to the regime were used up and the original motivations of the land reform lost their immediate relevance. The decreasing power of the Plan Organization, initially after the removal of its first head, Dr Ebtehaj,[7] in 1959 and then after the removal of management control in the late 1960s, left real authority with the executive ministries which had little, or in some instances no, interest in the direction of overall agricultural policy, especially as it touched upon the traditional farming areas. This situation was exacerbated by the division of responsibility for agriculture between a variety of ministries and agencies, between which there was frequent rivalry but rarely co-ordination.[8]

Thus within Iranian planning there was increasingly a lack of frank assessment of the real problems of agriculture and of effort to come to terms with its long-term needs. The country could afford financially to look for new solutions despite the clear fact that the traditional area was the largest sector of farmland, commanded the bulk of irrigation water supplies, occupied the best soils, had a monopoly of successful cultivation skills, and was a cheap employer of labour. The new projects after 1967 such as farm corporations and agribusinesses were divorced from Iranian reality and were poor imports in many cases. They were expensive but ineffective in themselves, and unfortunately carried the additional cost of causing far-reaching damage to the traditional village farming systems.

7.2 Planning for agriculture: agriculture in an oil-based economy

The changing role of agriculture throughout the contemporary period was brought about by policies of reform devised by the shah and his ministers in the Sixth of Bahman programme (Chapter 6), which, it has been noted, occurred outside the ambit of planning as such. A second extremely important influence on the pace and direction of change in Iran's economic structure arose from government policies towards the oil sector or rather failure to control it. The organic nature of the relationship between oil and other parts of the economy never seemed to be taken seriously by the authorities. Oil was perceived in two lights: first, as a factor in the political struggle against foreign, mainly British, influence and, second, as a source of revenue with which to fund the overseas costs of ordinary, defence and plan budgets.

The politicization of the oil issue in Iran was long-standing[9] and profound.[10] Belief in the rights of Iranian control of the industry and in the benefits that were to be gained by domestic management entirely overshadowed any reservations that might have existed concerning the adverse effects of oil export operations. Persian commentaries on the oil industry in the period 1920-70 rarely concede that *any* difficulties might arise other than those caused by foreign political or economic interference either directly or through the comprador bourgeoisie.[11] The best of the economic analyses portray the regional and structural defects of Iranian oil activities as if they were the products of foreign control rather than inherent in the nature of the petroleum economy. Mahdavy,[12] for example, in his excellent study of 1970, identifies many of the difficulties that beset oil economies, including those of the *rentier* state, the oil-producing enclave and the low grade of economic linkages between modern oil-exporting and traditional domestic subsistence economies. The implication of his findings is that the problems are as much a function of the external capitalist management of national Iranian resources as of in-built structural problems that apply equally in other oil-producing states such as Norway and the United Kingdom.

Katouzian in his analysis of the Iranian oil economy[13] adopted a less dialectical approach. He argued *inter alia* that Iran represented a particular type of oil economy — one with a substantial and presumably sustainable[14] oil-exporting sector but one which also ran in parallel with a large endowment in other, largely agricultural, resources.[15] He clearly outlined the potentially damaging effects of the oil sector on the parallel indigenous economy, though principally in the political context of mismanagement by the shah's regime. Other recent discussions have tended to follow much more politically oriented themes that are designed to attribute blame for the deleterious effects of oil on the economy to the corruption of the Iranian elite, the cupidity of the shah and his family, the neo-imperialist aspirations of the United States and similar factors.

It has been argued earlier in this study that the Iranian endowment in land and water resources is less than generous. It can also be shown that use of domestic resources in agriculture became relatively less and less efficient during the last fifty years (see section 7.4). In practice there was no absolute parallelism between the internal oil and non-oil economies but rather a dynamic of change that involved interactions between the two.[16] Indeed, a main conclusion of this study is that changes in economic structure saw an accelerating rate of decline in agriculture that was in direct proportion to the rise in importance of oil in the economy. While Iran was undoubtably an oil-exporting economy with significant non-oil contributions to its structure before 1973, by 1978 the situation had entirely changed. This is particularly so if conventional national accounting is ignored in favour of a 'wasting asset' analysis of the kind elaborated by Stauffer and Lennox.[17] Non-oil contributions to national income and employment measured in this way appear far less impressive than in conventional accounting procedures. A very large proportion, possibly 55-65 per cent on the basis of the Stauffer and Lennox model, of the Iranian economy originated in the oil sector in 1980 against 40-45 per cent a decade earlier, thus indicating a decisive structural change induced by oil effects in that period as experienced in the economies of other OPEC members.

Oil exploration and production were very small-scale in the early period. The influence of the oil industry on the rest of the economy was slow to develop and it was perhaps the imperceptible nature of the growth of petroleum activities that for so long disguised its true implications for the economy at large. This process was helped more than a little by the central position of oil on the political stage throughout the period from 1933 to the present.

Iran was the first Middle Eastern state to export petroleum on a significant scale following the discovery of commercial deposits of crude oil at Masjed-e Sulayman in 1908. Revenues from oil sales accrued to Iranian governments since that time, although their value was fairly limited until the 1950s. There were occasional fluctuations in the amount paid to the Iranian authorities on oil account. Military action in the Mediterranean area during the Second World War and the disturbed state of the Middle East over the Palestine issue after 1948 resulted in disruption of Iranian oil exports. Internal difficulties arose from time to time, including government-company disputes in the 1930s and 1950s. Nevertheless, despite the somewhat erratic income from oil exports, the clear overall tendency in the period before the 1979 revolution was for government revenues to increase as a function of both improved unit payments and/or an augmented volume of sales.

Oil revenues, denominated in pounds sterling, grew from £0.5 million immediately after the First World War to £4.3 million by 1939, by which time oil exports were the main source of foreign-exchange earnings. In the five years following the Second World War revenues all but doubled but collapsed in 1951-54 as a result of the Anglo-Iranian crisis (see Figure 7.1).

In many ways 1945 marked the end of the era during which

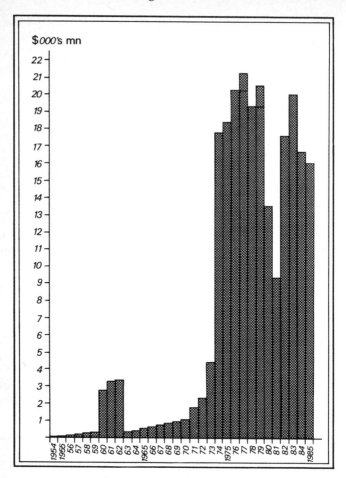

Figure 7.1. Oil revenues paid to Iran 1954-85.
Source: OPEC Secretariat, Vienna, *Annual Report,* various years.

oil revenues had only modest effects on the structure of the economy and the nature of economic development. Before then the limited income from oil was mainly channelled into defence and ordinary budgets or found its way into investment only in small-scale projects. In the years following the withdrawal of the Allied forces there was a discernible change in official attitudes, stimulated by a growing nationalist sentiment, resentment of dependence on foreign interests and a political need for economic modernization. The oil crisis of 1951-3 delayed the implementation of new policies but eventually the state moved to a position where oil revenues were specifically allocated to support planned development of the economy and from that time, which coincided with an expansion in oil revenues, there was an inevitable growth in the impact of oil on the indigenous economy. That the full meaning of this latter interdependence was rarely understood or taken into account lay at the heart of the failure of planning as an instrument of economic management and therefore was a root cause of the neglect of the country's long-term agricultural interests.

There was, none the less, a logic behind Iranian planning that tied in to the oil sector. The government found it convenient and necessary to appear to spread the benefits of oil income among the population. It was forced into such action by the very nature of the oil industry in which income accrued directly to it from sales, taxes, royalties and bonuses. The government had responsibility for allocating oil-derived funds across the economy, in effect providing access to imports of goods and services. Distribution of foreign-exchange resources for welfare purposes, as on the Kuwait model,[18] was an inadequate response for a country that considered itself rich in potential resources and was both more heavily populated and less likely to remain in a crude oil exporting role.[19]

With the passage of time the government elaborated a public philosophy towards economic development which suggested that the major part, comprising 70-80 per cent, of oil revenues should be used to develop the non-oil sectors of the economy. Depletion of oil reserves was to be compensated for by the creation of alternative productive assets that would ultimately provide a developed base for economic activity independent of petroleum. By this means, national and personal incomes could be sustained at current levels, even after oil exports had passed their zenith. Later, after the oil boom of 1973, this approach was interwoven with emphasis on adding value in Iran to hydrocarbons before their export. Throughout the period 1967-78 there was a certain urgency in the implementation of the ideas inherent in this development philosophy — as if the country had to take the positive economic tide at its flood lest the opportunity should never return. Economic planning was the chosen vehicle to enable the nation to make the transition from a backward oil-dependent economy to a more mature and broadly based structure.

Other forces were also at work. It can be argued that Iranian domestic economic policies and therefore the economic plans were formulated largely to achieve political objectives set by the shah. Economic and associated social development by the government were intended to improve the situation of the regime vis-à-vis its domestic constituency by generating support from the mass of people as living standards rose and the availability of welfare benefits grew. Growing national wealth functioned at the same time to enhance the shah's regional strength, by permitting him to finance an active foreign policy and to concentrate new resources on expanding his military forces. For long periods during the 1960s and 1970s it seemed that the shah was attempting to provide material and social benefits to the population at large so that he could justify the concentration of political powers in his own hands and the suppression of opposition. For the intelligentsia and the middle classes the shah made available both economic advancement and positions of authority within the expanding organs of state concerned with the economy, provided that they accepted exclusion from the central political arena. This set of policies, at times ill-defined and erratically implemented, demanded that the economy be kept at a high rate of growth so that satisfaction of a non-political kind could be given to a rapidly increasing population and a burgeoning group of entrepreneurs, traders, military cadres and others who could loosely be called 'middle-class'.

There is a sustainable correlation between the period in which the shah succeeded in maintaining policies for economic growth and improvements in the average incomes of ordinary Iranians, together with the provision of opportunities for position or profit for the middle classes, and the effective silencing of political criticism of his regime. The period 1964-75 stands out as one in which he was largely successful. With increasing effect from 1976 and especially from the Iranian year 1977/8, the government and the shah failed to maintain the link between the political and economic wings of their policies. Within less than two years the shah's regime had been swept away. There were other powerful domestic forces working against the shah in the period 1977-9 but the argument that there was an interrelationship between mass disillusionment on the economic front and the shah's downfall is none the less persuasive. Only when the framework of the planning process was abandoned by a government that appeared to behave as if a millennium of unlimited wealth had arrived and scarce real resources could be put in place overnight, did the country lapse into economic and political chaos.

Iran has fairly consistently passed through a cyclical process of economic expansion and decline on a number of occasions. The sudden flush of oil revenues accruing to the government from the oil price increase in 1973 did not reduce the economy's vulnerability to alternate swings from prosperity to depression as much as the politicians and planners hoped. Increasing oil wealth appeared to exaggerate the magnitude of the swing between boom and slump. The considerable advance in personal incomes of Iranians in 1973-6 was followed by a rapid fall on a scale that brought to the surface deep resentments and distress which were incapable of being managed by a government enmeshed by its own past incompetence in handling the eccentricities of an oil-based economy.

Any analysis of development in modern Iran is thus largely a review of the effects of oil revenues. When oil revenues were put at the service of a regulated economic plan for the benefit of the non-oil sectors of the economy, positive developments occurred, though not all were benign. The inherent bias of oil income in favour of centralization and urban growth and services was always pervasive even at modest levels of oil earnings. As soon as the utilization of oil revenues ceased to be carefully planned in relation to other parts of the economy or the growth of the oil sector became an end in itself, with scant regard to repercussions elsewhere, the country became a petroleum-dominated economy that differed only in scale from classic oil economies like Kuwait.

Iran was vulnerable to a collapse of planning authority and

unregulated expansion of the oil sector despite the early decision in the 1960s to allocate as much as 80 per cent of oil funds to its development plans. The problem arose because the Plan Organization represented an arm of government. Plan Organization staff formulated detailed and integrated development plans but these were always susceptible to amendment or rejection by the political authorities whenever they were found to be inappropriate or inconvenient. The status of the Plan Organization varied in importance. At times it was relatively powerful as in the period up to 1960. Later, in the years 1961-9, it functioned as an advisory and budgetary control device. After 1969 it was little more than a co-ordinator of projects devised by other state agencies with little ability to impose its own will. Nevertheless, up to 1973 the Plan Organization, for all its problems and shortcomings, influenced the overall scope, nature and direction of economic development very considerably. Even where major parts of the plans devised by it were put aside, they tended to re-emerge later for implementation in slightly different guise. The steel mill, for example, was included in all plans from 1955 but was officially adopted only in 1965 on an *ad hoc* basis following the shah's visit to the USSR and conclusion of the Irano-Soviet Joint Co-operation Agreement signed in 1966.

At the same time, the general pace of economic change and the sectoral pattern of growth set by the Plan Organization in the formal plans were taken by politicians as yardsticks of success or failure, although quite different mixes of projects from those originally envisaged by the Plan Organization were used to achieve this end. Development planning was, in consequence, an important formative influence on economic strategy until the Fifth Plan of 1973-8. And the role of agriculture as foreseen in the plans was invariably stated to be central. Why this did not occur in practice is examined in the following discussion.

7.3 Agriculture in the development plans

The First Seven-year Development Plan, begun in 1949, was researched and formulated mainly by American specialists, first Morrison-Knudsen in 1947 and, second, Overseas Consultants Inc. (OCI), who advised on detailed design and plan implementation.

During its first two years, the foreign consultants in the newly established Plan Organization were joined by a growing number of often young but strongly motivated Iranian staff under the command of Mr Ebtehaj. Oil revenues were expected to provide the greater part of foreign-exchange funding of the plan, supplemented by borrowing abroad. Immediate attention was required by the industries set up during the 1930s which had been allowed to deteriorate after the abdication of Reza Shah in 1941. The Plan Organization was made responsible for taking them over and improving their efficiency, but without much success. Badly run down by shortages of spare parts and a collapse of management during the Second World War, the industrial plants were in need of a more radical reorganization than could be provided by officials in Tehran.

Scarcely had the First Plan gained momentum when it was badly affected by events beyond its control in the oil sector, a bad omen for an institution designed to harness the benefits of oil exports. The *majles* refused to ratify the Supplemental Oil Agreement negotiated between the government and the Anglo-Iranian Oil Company. Anglo-Iranian was nationalized in 1951, an act which set off a dispute with the company and the British Government that in turn led to a cessation of most oil exports. Oil revenues fell to insignificant levels, exacerbating the shortage of foreign-exchange funds at the disposal of the Plan Organization brought about by the government's failure to raise other than what were considered inadequate American loans.[20]

The combination of economic shortages and a political crisis over the nationalization of the oil industry resulted in the plan all but ceasing to apply. It was proposed to spend 7,300 million rials, or approximately one quarter of all funds, on agriculture out of total allocations of 26,300 million rials during the course of the plan. In fact, expenditures by the Plan Organization on agriculture were only some 1,000 million rials and little was achieved. Indeed, the effects of the political unrest and the shortages brought about by lack of access to overseas markets appeared to leave agriculture weaker at the end than at the beginning of the plan period.

Following the end of the Anglo-Iranian oil crisis a new seven-year plan was drawn up and implementation began in 1955. On the whole it was little more than an investment programme for the public sector. Allocations to the Plan Organization were set at 84,000 million rials, of which 25,100 million, some 30 per cent, were for agriculture and water. By far the most significant investments were directed towards long-term projects in the water storage and irrigation sectors. Three large dam/water storage structures were to be constructed on the Sefid Rud, the Dez and Karaj rivers. Between them they were expected to absorb no less than 15,628 million rials, equivalent to almost two-thirds of the total funds available to agriculture as a whole over the plan period (see Figure 7.2).

In the early years of the plan foreign loans were used to prime the depleted financial reserves of the Plan Organization. In all, the Plan Organization drew $317.8 million in loans

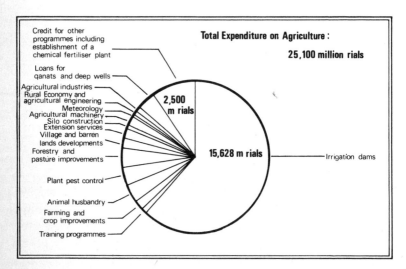

Figure 7.2. Second plan allocations to agriculture and water resources.
Source: Central Bank of Iran, *Annual Report,* various issues.

during the period, of which $168.3 million was from international agencies. A significant portion of this was from the World Bank to assist with the Dez dam project. It was hoped at the start of the period that 80 per cent of oil revenues would be allocated to the Plan Organization. This was never achieved, though ultimately approximately half of all oil income did find its way into Plan Organization coffers. Despite changes in targets for plan spending and erosion of the real value of allocations by rapidly rising inflation, agriculture and irrigation received 17,400 million rials during the Second Plan, representing about a quarter of all Plan Organization investments.

The benefits to agriculture were far less than was apparent from the pattern of spending by the Plan Organization. In addition to the near monopoly of resources by the inception of the three dam projects on their extremely narrow geographical bases, agricultural programmes were starved of funding from 1959 as the country experienced a worsening economic crisis. Virtually all the subsidiary budgets for agriculture were cut back as the effects of severe inflation and a deteriorating balance of payments position made themselves felt. Threats of land reform were an important depressant on private investment in agriculture beginning in 1960. By 1962 and the start of phase one of the reform entirely outside the aegis of the Plan Organization, the larger landlords were desisting from investment or making loans to peasants and there is evidence that some disinvestment was occurring as a small number of landlords razed orchards and plantations. In all, the performance of the second plan, despite its dramatic curtailment, was an improvement on that of the first but brought few immediate increments in agricultural output.

Optimism was not high when the Third Plan was brought in for the period September 1962 to March 1968. The plan was ratified at a level of 230,000 million rials, including 49,000 million for agriculture and irrigation, relegated, for the first time, in national priorities to second place behind transport and communications. As oil revenues expanded during the course of the Third Plan, there was a step by step increase in plan allocations. Actual receipts by the Plan Organization from oil revenues during the period were $1,913 million, close to 65 per cent of the total.

Disbursements on agriculture and irrigation by the Plan Organization amounted to 47,291 million rials, very little short of the original allocation. As in the second plan period, the water resources budget took up a sizeable amount of available investment funds as the dam projects continued towards conclusion after 1962/3. Land reform activities and the associated distribution of credits took up a substantial portion of financial resources (see Figure 7.3) as first and second phase implementation went on side by side. In effect, less than 30 per cent of all funds officially allocated in the plan to 'agriculture' were spent on productive investments outside the water and land reform budgets.

Trends in planning and official attitudes emerged during the Third Plan which were to signal changes in the designated role of agriculture. The contrast was apparent between the years before and after 1965. In the first period the government was engrossed in social and political reforms that revolved essentially around modification of the country's agrarian structure. Growth rates in the economy were low at that time and aspirations were confined by and large to gradual improvement and diversification of traditional economic activities. By contrast, after 1965, when initial arrangements were made with the USSR for construction of the Esfahan steel mill and associated machine tool plant at Arak, the orientation of development changed quite radically. The rate of investment picked up rapidly. Agreements were signed by government agencies with foreign interests for the construction of a series of major industrial projects such as petrochemicals and car assembly. In effect, the government launched itself on a drive to industrialization in the mid-1960s that brought instant success by way of modernization, job-creation and a surge in the rate of economic growth. The improvement in the economy during the period 1965-8 was the more marked in relation to the poor performance of the latter part of the Second Plan. In the Third Plan taken as a whole expenditures were far larger than in the preceding seven-year plan and there was far less inflation. Easy success with the *ad hoc* adoption of development projects in industry led the government away from formal planning

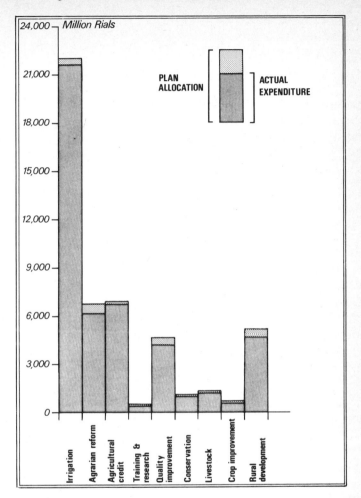

Figure 7.3. Third plan allocations to agriculture and water resources. Source: Ibid.

routines and notions of balanced growth based on agriculture.

The Fourth Plan began with the country in a much stronger position than at any time in the past. The regime appeared firmly in political control. Oil revenues were rising quite rapidly in response to the government's consistent and generally successful policy of increasing the volume of oil exports.

Throughout the 1960s Iran achieved a significantly better rate of incremental growth in output than most of its partners in OPEC.[21] At the same time, the acceleration in economic development experienced in the third plan period gave the new programme a useful early impetus, since many of the investments started in the Third Plan were carried over.

The Esfahan steel mill, the Arak and Tabriz machine tool plants, the Bandar Shahpur (Bandar Khomeini), Abadan and Kharg Island petrochemical projects, the aluminium plant at Arak and others ensured that industrial expansion received highest priority in the plan period, taking 22.3 per cent of allocations. Agriculture was relegated to fourth largest recipient of funds with 46,700 million rials, less than 10 per cent of the total. Water was separated from agriculture for purposes of fund allotment in the Fourth Plan. Estimated expenditures on water were budgeted at 43,300 million rials, though a large proportion of this was for urban and industrial water supply schemes. Investments in water from both the fourth and earlier plans became increasingly used by the urban industrial sector either directly or for the generation of hydroelectric power.

Of the funds available to agricultural development, which varied as increasing oil revenues permitted an augmenting of the plan's financial resources, notably in 1971, the largest allocation was earmarked for the creation and support of farm corporations and grants to commercial farmers. Agribusiness projects which assumed importance as the plan progressed were initially incorporated in the budgets of the Ministry of Water and Power and it is not clear from government accounts to what extent funds were switched to/from agriculture for this purpose. A large fund of over 6,000 million rials was set aside for the protection of natural resources, while agricultural research and regional studies — a vital area given the poor state of the data base[22] — took 4,000 million rials (see Figure 7.4). Once land reform, farm corporations, conservation and other non-basic agricultural activities in the hands of the state were taken into account, the poverty of the budget in support of production in the traditional sector was extremely, almost derisively, small, with a mere 25,500 million rials at the most generous estimate from a national budget originally set at some

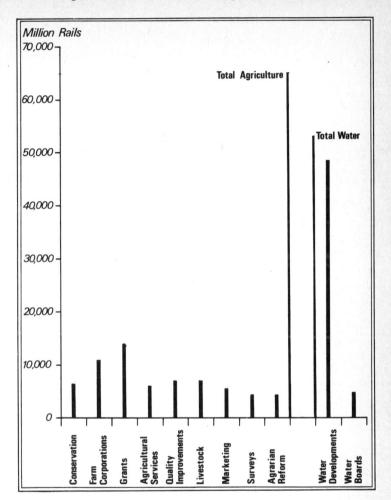

Figure 7.4. Fourth plan allocations to agriculture and water resources.
Source: Ibid.

554,500 million rials.

It was not surprising in these changed circumstances that the rates of growth achieved in agriculture were less than forecast. The target was dropped during the course of the plan from 5.0 to 4.4 per cent and eventually worked out, according to official sources, at 3.9 per cent in real terms. Continuing activities in the land reform under phase three acted as a brake on investment in the traditional farming area, while progress in the modern sector in reclamation of land under irrigation and dry farming was slower than expected. Only 400,000 hectares of new irrigated land and 150,000 hectares of dryland were brought into use despite the completion during the plan of the Aras, Kurosh Kabir (Zarinehrud), Shahpur I (Mahabad), Voshmgir, Shah Abbas Kabir (Karun I) and Dariush Kabir (Dorudzan, Shiraz) dams. While there were gains made in crop production over the plan period, they were comparatively small-scale when measured against the investments made.[23] Improvements were unable to keep up with the growth in population, and lagged severely behind strong growth in domestic demand for more and better-quality foodstuffs. An index of this problem was provided by the increase in agricultural imports which ran at $1,100 million or some 11.5 per cent of all imports for the plan period as a whole, a trend that foreshadowed the general swing into agricultural imports and external food dependency in subsequent years.

The condition of agriculture went into perceptible decline during the fourth plan period not so much because of the plan itself but from the insatiable desire of government departments other than the Plan Organization to intervene in its affairs at the structural level. New institutions and rising investment in them were inadequate to offset deterioration in the traditional farming areas. Agricultural activities slumped as contributors to national income, though they lost importance as employers somewhat less rapidly in the face of what appeared to be spectacular growth in industry and commerce. The Fifth Plan did nothing to save the situation. Exasperation among the country's economic managers with the intractability of agriculture's problems, increasing rewards from the industrialization programme, and (even before October 1973) a strongly rising income from oil exports meant that few tears were shed in official quarters over the abandonment of village-level agriculture to a minor role. In parallel with this implicit view of agriculture on the part of the government went a belief that state intervention in the sector through farm corporations, production co-operatives and agribusinesses would generate some growth in output to offset the failures of traditional farming.

At first sight agriculture was not neglected in the Fifth Plan, 1973/4-77/8. Of total investments of 2,848,100 million rials scheduled under the revised plan of August 1974, 296.74 million rials was allocated to agriculture. In effect, the government transferred very little of the new funds available after the oil boom to agriculture. The projects for the agricultural chapter of the plan were predictable. Large sums, totalling in all 369,440 million rials were allotted to credits and investment for the public and commercial sectors (see Figure 7.5). Consolidation of the rural population into nominated growth poles went hand in hand with the establishment of more state-run production co-operatives and farm corporations on some 900,000 hectares within the traditional village farming area. Approximately 400,000 hectares of so-called virgin land, though rarely outside the traditional farming economy, was to be brought into agribusiness ventures.

The economic prescription for agriculture under the Fifth Plan was designed to increase the rates of investment and productivity where the easiest short-term gains could be obtained. To assist planned investment to achieve increased productivity per man, those involved in the process were to be brought together in large management units and all others were to be ignored. There was already abundant evidence in 1972-4 to show that rural migration was increasing.[24] It must be assumed that the planners deliberately intended to set in motion a further acceleration in movement to the towns as a means of adding to the pool of labour available in the urban areas so that shortages should not inhibit expansion of industrial and related employment there. The transfer of labour out of agriculture and into new activities was, in principle, a reasonable economic response to the experiences of the fourth plan period. At that time, however, the flow of oil revenues was comparatively limited and enabled their absorption without serious side-effects.[25] The Fifth Plan was implemented in entirely different circumstances on a tide of rapidly increasing oil revenues on a scale never before experienced. Unpre-

174 The Neglected Garden

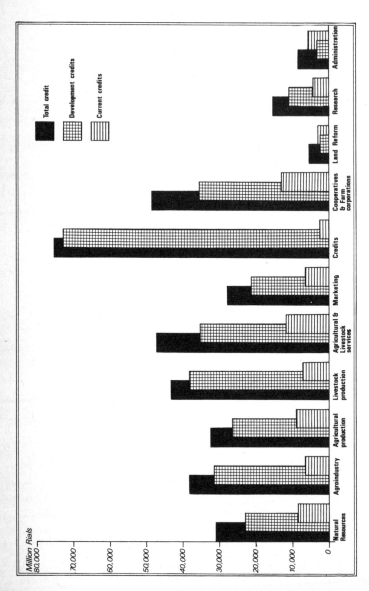

Figure 7.5. Fifth plan allocations to agriculture and water resources. Source: Ibid.

cedented urban demand for labour in the years after 1973 came to bear on a rural workforce that had been systematically uprooted from its established economic role and subjected to appreciable social disruption. In such a situation resistance to the urban demand for labour was all but negligible. Unfortunately the losses in the rural workforce were also losses in the reservoir of skills available for agriculture in some villages on a scale that was all but fatal for the farming communities.

Implementation of the Fifth Plan was attempted despite the externally induced changes in circumstances. New projects were also adopted through a series of bilateral agreements with the industrialized countries based on oil-for-goods exchanges. These were for the most part in addition to the schemes within the plan. At the same time, the armaments acquisition programme was advanced so that the defence budget took 1,968,700 million rials or approximately one quarter of all allocations under the revised plan. These and other increases meant that government expenditures rose precipitately. Meanwhile, the availability of domestic real resources of manpower and physical assets grew but slowly. Attempts to import skilled and manual labour to fill the widening gap between demand and supply was only partially successful, since the process carried new costs and supply problems in housing and other services. The physical and commercial infrastructures were totally inadequate to carry the burdens imposed by unrestrained government spending on projects. There were serious bottlenecks at the ports, in the internal transport system and with basic commodities such as cement and other construction materials. Inflation rose rapidly during the period to some 27 per cent according to very conservative estimates. It was augmented by the effects of imported inflation, associated with the rising prices of goods and services bought from abroad with the country's oil income. Increases in the allocation of funds to development were largely lost through the problems of malabsorption during the period of intensive spending that affected the beginning of the fifth plan period. The economic boom was characterized by one cabinet minister in conversation with the author as a year of illusion in which grandiose plans were conceived but little more achieved than had been planned before the boom began. Genuine gains during the boom years as a whole were mainly in the form of consumption of imported goods by both the state and the private sector.

Agriculture was spared the direct impact of a vast increase in expenditure under the revised Fifth Plan. As noted, its allocations were given only modest additions in 1974, but the objectives for agriculture were set at high levels. A growth rate of 7 per cent was proposed and a steady move towards food self-sufficiency was expected to arise from extremely rapid increments in physical output. The means towards achieving these goals were less clear-cut than for many other sectors. A further push into state-controlled large-scale farm units, a greater measure of mechanization and an improved farm infrastructure were prescribed. The plan carried little real conviction since it held to the pattern of modernization that had failed the country during the preceding Fourth Plan, and contained no cures for the deepening difficulties affecting the traditional farming areas. Experience in the later years of the Fourth Plan had indicated that rapid growth in the urban economy inflicted severe costs by way of adverse trends in income gravities and migration on the rural sector. Yet nothing was done in the subsequent plan, which was far more geared towards developments in urban areas, to offset the adverse effects on agriculture. There were some tools at the government's disposal, including channelling of funds into investment in the rural districts, adjusting the price mechanism for agricultural commodities to favour producers, or limiting agricultural imports, but none was provided for in the plan and none ultimately deployed.

Implementation of the Fifth Plan suffered from the chaos and waste brought on by the problems of government spending at rates that could not be successfully absorbed. This situation was compounded by a marked deterioration in the real value of oil earnings and a failure of oil revenues to live up to earlier expectations. In 1975/6 the government attempted to rein in spending and bring the plan back to the original guidelines of the pre-boom period. Activities of the private sector were cut back by a reduction in the flow of credits and through political steps against the main entrepreneurial families and industrial groups. A Public Ownership Law was introduced on 24 June 1975 which required 99 per cent of the equity of public companies and 49 per cent of private industrial companies to be sold off by their managements. An anti-profiteering campaign was also directed against private industries and the retail trades in July 1975. These measures proved inadequate to cope with problems of shortages and inflation at home allied to the deteriorating purchasing power of oil revenues. Dr Amuzegar was appointed Prime Minister in 1977. He endeavoured to halt burgeoning government expenditures by introducing an economic stabilization programme which curtailed spending on all but vital projects. By the last year of the Fifth Plan inflation began to fall but at the cost of declining real incomes and increasing unemployment.

The performance of agriculture during this period was determined by the mainly negative effects of activities elsewhere in the economy. Production was reported by the government agencies to have risen during the plan period, with notable gains claimed in cereals. There was also some diversification of crops as maize and oil seed cultivation spread gradually throughout the major farming regions. Official figures tend, however, to disguise losses of output of vegetables, orchard crops and other items where intensive horticultural cropping was abandoned because of labour shortages. The picture portrayed by the statistics is also put into question by the deteriorating output from agro-industrial enterprises. Their performance was so poor that many of the foreign managements pulled out and they were forced back into state ownership and kept in being only by subsidies.[26] The farm corporations had little more success and even a showpiece unit such as the Arya Mehr Corporation near Esfahan was reliant upon external aid.[27] The catalogue of failure was summed up by the growing value and volume of imports, which reached $2,550 million in 1977/8 alone.

In the traditional village areas there was considerable change. Up to two and a half million rural migrants are estimated to have moved from the countryside during the Fifth Plan period.[28] Wage rates in agriculture fell dramatically vis-à-vis urban occupations, with farmers receiving only 30 per cent of the average national wage by 1975/76. Returns on private investment in agriculture were also relatively poor compared with opportunities elsewhere so that the best entrepreneurial skills were tempted away. This problem, already pernicious as a result of the land reform programmes, was exacerbated after the anti-profiteering and ownership laws of 1975 by growing fears of arbitrary measures against capital by the shah's regime.

From the ruins of the Fifth Plan the government under Dr Amuzegar began to rebuild a new policy towards agriculture in 1977/8 to be incorporated in the Sixth Plan (1987/83). The plans for agriculture foreshadowed remarkably closely those ultimately adopted by the revolutionary regime after 1979 in so far as they looked forward to a much greater degree of self-sufficiency in foodstuffs, eventually to reach 80 per cent of total internal demand. Less heartening was the belief in the outline 1978-83 plan that more consolidation of farmland would create conditions in which greater self-sufficiency could be attained.

7.4 Agriculture and structural change

The history of change in agriculture's role in the structure of the national economy during recent decades was one of accelerating comparative decline (see Figure 7.6). Despite uncertainty over the statistics, the general long-term trend was clear. Petroleum was the main influence modifying national income, though not on an appreciable scale until after 1954/5. Before then the impact of the oil sector was limited by the relatively limited flow of oil revenues (e.g. £1.1 million in 1925, £2.2 million in 1935 and £5.6 million in 1954) and by the almost total disappearance of oil income for 1951-53 as a result of the Anglo-Iranian dispute. During the Second Plan period oil became of growing importance. Revenues went up by 29.4 per cent and oil added some 7 per cent to Gross National Income over the period according to the Plan Organization. Low rates of growth in the agricultural sector in comparison to movements in other areas of the economy prevailed throughout the 1960s and 1970s, thereby diminishing its importance at an accelerating rate. During the period 1963/4-1967/8 agriculture

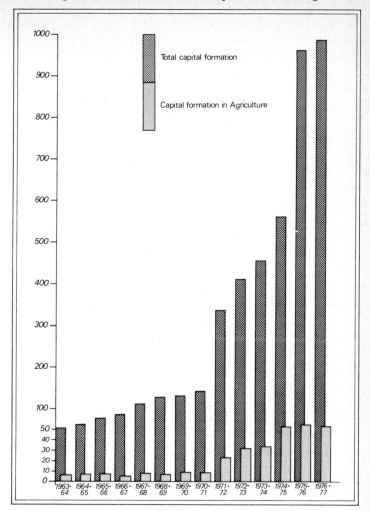

Figure 7.6. Capital formation in agriculture 1963/4 to 1976/7.
Source: Central Bank of Iran, *Annual Report*, various years.

advanced at an annual average of 3.6 per cent and did only slightly better in the following 1968/9-1972/3 period at an average of 3.9 per cent. For the same two periods Gross National Product rose by annual averages of 9.5 and 11.8 per cent, respectively.

The first useable estimates for national income were made

for 1958/9 and suggested that agriculture comprised 25.8 per cent of Gross Domestic Product (though the Statistical Bureau of the Ministry of Economy suggested approximately 35 per cent for the same year). Accepting the frailty of the available statistics, there was none the less a perceptible downward trend in agriculture's contribution to national wealth. By the close of the 1960s it had fallen to 22.0 per cent. The rate of deterioration was very fast during the Fourth Plan so that by 1972/3 it contributed 15.7 per cent to GDP, and with the powerful impact of the oil boom it had sunk to less than 10 per cent of GDP by 1977/8. Excluding the *direct* contributions of oil to GDP, it appeared that agriculture made up only about 15 per cent of the country's non-oil production of goods and services by the end of the Fifth Plan, 1977/78 (see Figure 7.7).

Changes of this order were also reflected in the structure of the labour force. Statistical data here were less reliable but it was suggested by the Iranian Statistical Centre that the workforce in agriculture stood at some 3.33 million in 1956, 3.67

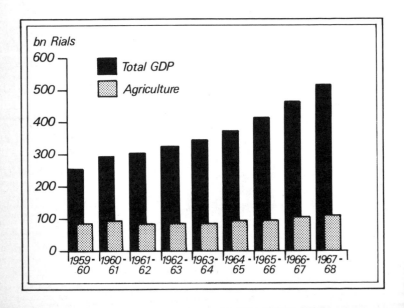

Figure 7.7. Trends in agriculture's contribution to GDP.

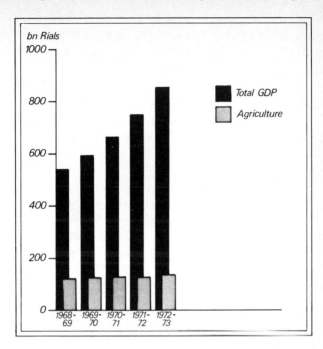

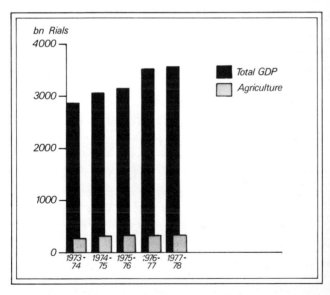

Source: Central Bank of Iran, *Annual Report,* various years.

million in 1962 and rose to a peak of 3.80 million in 1968. From that point there was a steady decline to 3.57 million in 1975. Trends after that year are less certain. Migration to the towns may temporarily have lessened so that numbers in the countryside picked up slightly in the final years of the Fifth Plan. It was not entirely apparent to what extent foreign labour took the place of Iranians in agriculture during the boom years of the 1970s. Certainly, large numbers of Afghans and some Pakistanis, in Varamin for example, were observed by the author while travelling in the Iranian countryside in 1978. That a mass transfer of population from rural areas to the towns took place at that time is indisputable. The latest census data suggest that most of the movements were permanent. The proportion of the labour force in agriculture, which was reported by the 1956 census at 56.6 per cent, fell to 49.0 per cent by 1966 and 24.5 per cent by 1976.

The economic development plans which seemed to give such emphasis to agricultural development actually allocated less than generous funds to agriculture, often failed to ensure that these funds were disbursed, and did not achieve a high level of investment for the sector. The poor showing of the Plan Organization in disbursing funds has been commented upon above. The position for capital formation in agriculture as a proxy for total investment indicates both the difficulties endemic to the sector and the particular problems that affected the Iranian administration.

It is always a problem that the funding of the running costs of agriculture is high and that allocations by governments must take account of high rates of recurrent costs. More than industry, agriculture presents obstacles to a clear definition of capital and recurrent costs. In Iran these issues were compounded by a government which generally took 'development' to mean modifications of the agrarian structure, as is apparent from the earlier discussion of the land reform (Chapter 6). The continuing nature of structural change meant that financial resources allocated to agriculture were constantly taken up in activities designed in theory to remove obstacles to investment but which themselves diminished the prospects for asset-creation, not least by consuming available resources in administration costs. Reform also discouraged investment by the private sector within the traditional village system, especially once the farm corporations were announced.

Indicative of the extremely poor record of agricultural investment over the medium term are the data on capital formation published from time to time by the Central Bank. Using constant price series, investment through the Third and much of the Fourth Plans remained at modest levels (Figure 7.6). At the beginning of the Third Plan approximately 10 per cent of capital formation took place in agriculture but the relative decline after that date was marked by two years, 1966/7 and 1976/7, when the situation was particularly disappointing. The low rates of investment indicated that the deterioration in agriculture was not simply a question of the relative growth of oil, industry and services but also one of severe neglect. Such low investment levels were an active reason for the undermining of the sector and a contributory explanation of the sluggish growth rates experienced in an area already adversely affected by preoccupation with land reform. The irony of the situation was emphasised by the demise in the late 1970s of the agro-industrial units and other modern farming structures which had absorbed a great deal of the investment that had been made by the state agencies.

7.5 Agricultural planning and regional change

The rapid relative decline in agriculture had important implications for regional change. First, there was a pronounced discontinuity in the rural-urban population balance as a result of the accelerated development activities of the 1960s. The number, size and function of towns and cities multiplied many times[29] especially in the zone of the fertile crescent running from Lurestan through Kurdestan and Azarbayjan to Gilan, Mazandaran and Gorgan on the Caspian coastal plain, where urban centres grew in proportion to the wealth in production and population of their hinterlands. Polarization on demographic grounds was at its strongest between villages and major cities, the latter expanding very rapidly and the former generally with a static level of population.

The government made efforts within the development plans to reduce the growing disparities between the cities and the countryside, but with little success. Tehran increased its primacy together with centres such as Esfahan, Mashhad, Tabriz and Shiraz. At the same time as urban areas were expanding so rapidly, the rural population as a whole, both village and nomadic, remained virtually unchanged. There were absolute declines in the village-based populations in Gilan, West Azarbayjan and Esfahan during the inter-censal period, though boundary alterations and statistical errors mean that such falls cannot necessarily be taken as harbingers of permanent structural changes. Urbanization ran at an estimated 4.9 per cent each year in the period 1960-70 and 5.0 per cent in the following decade, so that the proportion of the population left in rural areas fell from 66 per cent in 1960 to 49 per cent in 1980.[30] The new regional pattern of population distribution reflected not only the obvious paramountcy of the urban over the rural economy but also the growing importance of the more climatically favoured and natural-resource-endowed areas of the country over others. Within that formula Esfahan, Gilan/Mazandaran/Gorgan and Khuzestan stood out as areas of positive correlation between share of national income, attraction for investment and strong endowment with natural resources.

Tehran, as the capital of what was a centralizing regime, benefited considerably from the available investment and enjoyed a disproportionate share of national income. The desert areas of the centre, the south-east and the east did worst as regards the existing share of national income and in attracting new investment. Azarbayjan had a modest share in national income but appeared in the early 1970s to be gaining an improved proportion of capital investment. The desert areas, in particular, appeared to be poor and destined to become poorer even before they were struck by the negative effects of the oil boom. That the poorer rural areas showed least income inequality between households[31] was small compensation for those in agriculture in the declining areas.

Intensifying regional inequalities which inevitably affected the prosperity of agriculture within the individual regions arose from the stress on industrialization and the growing importance of the petroleum industry. The granting of industrial licences consistently favoured Tehran and the Central Province, which always took more than half and often two-thirds of all new private-sector projects. Kerman, Esfahan, Khuzestan and Mazandaran were a second tier of beneficiaries, with the other provinces receiving insignificant attention. Industrial investment by the government tended to set the pattern followed by the private sector. During the 1970s, for example, approximately 30 per cent of all fixed investment by the state was directed into the petrochemical industry in the oilfields region of Ahwaz, Bandar Shahpur (Khomeini) and Abadan. Development of government-owned iron and steel plants favoured Esfahan (steel mill), Kerman (for raw materials for the steel mill) and Ahwaz (rolling mills), and reinforced the regional imbalance. The rapid growth of the oil industry in south-western Iran, together with its linkages via domestic oil refineries at Tehran, Tabriz, Kermanshah (Bakhtaran) and Shiraz, added to the general economic expansion in those areas and diminished the relative importance of the unindustrialized provinces.

The planners were not entirely indifferent to the differential regional effects of economic development activities. From the first plans lip-service was paid to the needs of the regions and the disadvantages of the monopoly of Tehran and the oilfield enclave over financial and other resources. In fact, far more development in the provinces came from the project-oriented nature of the early plans than from the specific provisions within what might loosely be classified as regional planning. In reality, regional planning simply occurred in the project areas of schemes included in the plan for reasons unassociated with the planning of the region. In the Third Plan concern was almost exclusively with the three dam projects on the Rivers Dez, Karaj and Sefid Rud, the benefits of which were automatically tied to the geographical areas served by them, especially before the creation of the national electricity grid. In the Fourth Plan the start-up of work on petrochemical, steel, machine tool and aluminium projects had strong effects within their local regions, though an attempt was also made at the time, 1968-73, to elaborate an integrated policy for regional development.

The government hoped that it could force investment away from Tehran to those provinces where the presence of dense population, above-average literacy and training, sound agricultural hinterlands and the principal items of physical infrastructure offered a basis for development. For the very large area of the country that lay to the east of meridian 53, which was, with the notable exceptions of Mashhad and Kerman, mainly excluded from priority development, there was effectively no regional development planning other than in welfare. As part of its programme for the regionalization of development, the Ministry of Economy obstructed the grant of licences for private-sector industrial schemes that were within a 120-km radius of Tehran, and gave modest tax and other incentives for companies to set up their businesses in designated growth poles which included Esfahan, Ahwaz, Tabriz and Arak in the first phase and Shiraz, Bandar Shahpur (Khomeini) and Qazvin in the second phase of policy. Rasht, Mashhad and Kermanshah were nominated as sites for later developments.

A large measure of success was achieved in establishing sites under the first two phases by the end of the fourth plan period. A secondary tier of industrial estates for small industries was also set up at that time as separate but underpinning projects for the major industrial poles. Small-scale industrial estates were planned for Ahwaz, Qazvin, Tabriz, Esfahan, Shiraz, Arak, Kermanshah and Rasht, among others. There was a remarkable degree of industrial expansion at the estates set up under the Fourth Plan, though rarely was there any linkage between the individual projects and the local agricultural sector on the estates that the author visited during the early 1970s.

Policies under the Fifth Plan were less consistent than in previous plans. The *ad hoc* adoption of projects in the period of the oil boom following 1973 and the considerable increases in the number of schemes put in hand swamped the organizations directing regional development. The plan itself called for state-sponsored investment to be channelled into towns and villages with populations of more than 5,000 persons, and withdrawal of official funds from small settlements of less than 250 persons. The damaging aspects of this programme were remarked on earlier, but it is interesting to note that the provisions of the Fifth Plan would have led to the integration of agriculture into regional development for the first time since the inception of the formal state economic plans in the 1950s. In the event, the plan was all but abandoned. Regional development followed rather than led project adoption once again. Agriculture was neglected as a component within the nexus that passed for regional development planning, since so few of the individual projects taken on during the years of the boom were agricultural in character.

7.6 Conclusions

The rural areas of the country underwent thoroughgoing changes in the post-1945 period. They began slowly in what was then to all intents and purposes an agricultural economy and a village society. Rural areas were affected by two pressures which grew to irresistible proportions in the 1960s: first, the land reform and its subsequent programmes which restructured land ownership on a continuing, but not consistent, basis between 1960 and 1978: second, by the economic modernization set in train by the central government and articulated through development plans and contemporaneous development projects. Interventions in agriculture by the central authorities were possible on an expanding scale because the oil revenues at their disposal increased fairly steadily during the 1960s.

The rate of increase in oil revenues in the period before 1972 was sufficient to fund economic expansion at a pace which enabled the country to absorb a high ratio of funds in productive enterprises. Objectives laid down for, and management of, the conversion of oil earnings to domestic use were generally laudable within the existing economic context. There were, however, costs other than financial ones to the developments undertaken. The indigenous economy was subverted by oil-dependence even at that stage. Iran was gradually but inexorably becoming far more of an oil-based economy than was shown in conventional presentations of the national accounts. There was also a demonstrable deterioration in the cohesion and relative prosperity of much of the agricultural community which arose partly from its devaluation by absolute additions to the national income from oil activities and partly from the disproportionately greater beneficial effect of growth in petroleum revenues in the towns.

This chapter's study of the economic plans devised by various Iranian Governments suggests that they were variable in their impact between different sectors of the economy and various regions of the country. The gains attributable to economic planning in its broadest context were most visible in the central services of the capital city, in urban industry and in the expansion of oil processing in the oilfield zone. In contrast, rural society and the agricultural economy were dealt with for much of the post-1945 period outside the framework of the plans as a result of continuing land reform under separate legislation and a different administration.

Where agriculture was included within the economic plans, resources were less than generously allocated to it. Investments in large-scale water storage and supply systems and modern commercial agriculture were often sizeable, but allotments to the much larger traditional farming area were miserly and seemed for long periods to be less than basic care and maintenance demanded.

Failures in the sectoral plans for agriculture were not mitigated by efforts at integrated regional development. The urban areas flourished while the countryside stagnated. Regional planning ultimately brought the threat of a mass restructuring of rural settlements under the growth-poles policies of the Fifth Plan period, from which the villages were saved only by the mismanagement of the oil boom and the abandonment of the formal plan. This reprieve was short-lived. The overwhelming influence of the boom years 1973-6 reacting with the rural economy in an unplanned and catastrophic way induced rising migration to the towns and the syphoning-off of the agricultural labour force. In summary, not only did the plans as a whole neglect the country's agricultural interests but they also broke down so completely in 1973/4 that they exposed the countryside to the rigours of a sustained urban oil boom against which its defences had been systematically weakened. Predictably, the outcome was the introduction of severe and possibly irreparable damage in large areas that had been under cultivation in traditional village agriculture.

8 Agriculture, Revolution and the Rural Community

8.1 Agriculture in the prospectus of the revolution

The root causes of the Iranian revolution remained the subject of intense debate for a considerable period after 1979. Quite properly, many of the discussions in the Western literature concentrated on the role of Islam.[1] Others have proposed political causations within either a marxist analysis[2] or an indigenous socialist framework.[3] Several observers suggested that economic factors played an important part in bringing down the shah's regime.[4] The complex interplay of these and other factors means that none can be put aside lightly. But, whichever explanation or combination of elements is chosen as having been involved in bringing about the overthrow of the former regime, the position of agriculture and the rural community will inevitably be an integral part of the argument.[5] A significant motivation for the Iranian religious leadership to stiffen their resistance to the shah was the land reform, which in its later phases adversely affected the status of land and water holdings constituted as endowments for Islamic charities by moving control of them firmly into the hands of the state. Although Ayatollah Khomeini himself did not make the reform of shrine lands a major issue in his clash with the government in 1963,[6] other members of the *'ulama* at that time did attack the regime for breaches of Islamic law. The land reform and other parts of the shah's 'white revolution' remained as continuing sources of friction between the government and many of those within the religious establishment.

On a different level, the state of agriculture caused affront to the *'ulama*. The decline in the importance of farming activities vis-à-vis other sectors of the economy automatically reduced the natural constituency of the *mollas* in the rural areas. Not only was the volume of donations in cash and kind reduced in relative terms as population moved away from the villages but the type of traditional society that appeared as the ideal environment for the Islamic way of life was also diminished by the changes introduced by the shah's governments in the years after 1963. Taken a step further, it could be argued that village life and its traditional virtues were destroyed by the shah as sacrifices to his new society and economy which seemed to the *'ulama* and many others in Iran at that time to be a mere copy of Western materialism. In so far as modernization was equated with Westernization and the anti-Islamic values that went with it, the Iranian and Islamic cultures associated with rural life and agriculture were perceived as reservoirs where the alternative value system to that imported by the shah was preserved. The failure of the shah's regime in its policies towards agriculture and its neglect of the rural communities in search of some elusive vision of a 'great civilization'[7] were taken as an affront and exploited as a political tool by those clerics already ill-disposed to the shah's aspirations.[8] Khomeini's speech in 1964 outside his house in Qom shortly before going into exile summarized his attitude to the matter, 'I am aware of the hunger of our people and the distorted state of our agrarian economy'. In such circumstances it was not surprising that the future role of agriculture and of the population still resident in the villages should play a prominent part in the prospectus of the revolutionaries.

In fact, most if not all of those groups who participated in the revolution incorporated in their manifestos some mention of plans for agriculture once the shah had been displaced. The prescriptions were varied. Even amongst those surrounding Ayatollah Khomeini there were conflicting views. But the stress laid on agriculture was strong if not universal among the revolutionaries. There was unanimity in condemnation of the maltreatment of the peasantry and the failure of government policies to bring about a positive development of farming. The shah's land reforms were vilified.[9] When it came to proposals for the salvation of agriculture, however, the various sections within the opposition movement were either at a loss or deeply at odds with one another. Positive programmes for improving productivity and standards of living in the farming communities were sadly lacking once the negative rhetoric against previous policies was stripped away.

The lack of real dedication to agriculture on the part of the revolution is perhaps best explained by its urban nature. Although Ayatollah Khomeini claimed that he was aware of conditions in even the most remote Iranian villages[10] and always spoke for all of Islamic society in the country, the geographical locus of revolutionary activity was almost exclusively in the cities and larger towns. The events of 1978 were aptly called an urban revolution.[11] When the rural areas were affected by agitation against the shah and his government, the agents behind these activities were often returning migrants importing their newly acquired urban ideas back into the villages.[12] In effect, while the villages were not entirely passive, neither were they in any way active components of the revolutionary process in 1978-79. Representation of the genuinely rural and farming interests in the institutions that appeared immediately following the revolution was thus very weak, rather as it was before 1979. Concern with agriculture was more powerfully weighted towards rhetoric and political aims than towards any improvement in the position of the rural community.

The most obvious concern arose in respect of the need for self-sufficiency and an end to external dependence — particularly dependence on the imperialist powers. Ayatollah Khomeini saw that US sanctions against Iran were a great risk and promised that 'we can... content ourselves with the barley and corn that we sow on our own land'.[13] In a major survey of the road to self-sufficiency in the *Ayandegan* newspaper of 22 May 1979, it was suggested that the first steps towards minimization of the country's economic dependence should be freedom of operation of the laws of supply and demand within the domestic market and improvement in the marketing of agricultural goods. At the selfsame time other groups within the government, not least those in the Ministry of Agriculture, were preparing to establish a system for control of farming activities that was entirely centralized in its management.

In general, therefore, once the regime came to power it was remarkably unprepared to take the situation in hand, despite the recognition of the great importance of agriculture. The first Minister of Agriculture, Ali Mohammad Izadi, suggested that success or failure in agricultural policy could make or break the revolution.[14] He had few remedies, however, for the problems affecting the sector. The measures taken to stimulate farming included some emergency grants of interest-free loans and the release from their jobs of 33,000 ministry staff so that they could join in farming activities. For the rest, the authorities suggested longer-term involvement in improving educational, health and transport facilities in rural areas but no well-thought-out programme of development. By the middle of 1979 the ministry was engaged in the complex issue of land ownership once again, with seizures of land by peasants and other opportunists causing it considerable problems. Its ability to construct a thorough development strategy was overtaken by the management of the considerable day-to-day problems surrounding the administration of domestic agriculture and the country's food import needs.

Two issues appeared to preoccupy the Ministry of Agriculture in the immediate wake of the revolution — the land ownership question and the low level of financial allocations to agriculture within the development budget. Changes in land distribution during the revolution were appreciable. Peasants took land from landlord holdings, hoping that the authorities would connive at their actions. Corrupt officials exploited the opportunity of the chaos during the revolution to take over land in their own names. Other areas were seized from former members of the shah's government and projects financed by foreign capital such as those in the Dez irrigation scheme were threatened with sequestration. The ministry's deep concern about the land question was bound to bring frustrations. Minister Izadi became engrossed in the problem of land reform but lacked the political strength to carry his proposals through the weak cabinet of Engineer Bazargan. He and a number of his successors faced the insuperable obstacle of a rejection by the religious leaders of arbitrary action against private ownership on the grounds that it was un-Islamic. A great deal of effort and time was wasted by the ministry in confronting the religious establishment over land reform, despite the clear supremacy of the Council of Guardians of the Revolution in judging issues of this kind and the known ambivalence of Ayatollah Khomeini on the matter.

The failure of the ministry's main policy was allied to a continuing inability to command a large portion of the development budget.[15] This combination left the ministry in a weak position, and was a partial explanation for the extremely dismal performance of policies towards farming and the rural infrastructure which characterized much of the post-revolutionary period.

Only one other political group put into practice its ideas on agriculture, the revolutionary guerrilla group, the Mojahedin-e Khalq,[16] in the northern cotton-growing area of Gorgan. Here there was an attempt at instant land reform by the systematic seizure of land from the larger landowners and absentee landlords — amounting to 5,448 hectares. Local committees controlled by the left-wing guerrillas were reported to have enforced collection of up to 20 per cent of the shares in the harvest of farmers[17] in the north. Among the interesting policies introduced by the local committees in the Turkoman areas was an attempt to oust landowners who were not Turkomans and to enforce local autonomy in farm administration. In the event, the central authorities crushed the Mojahedin committees and with them the Turkoman minority. The cost of the enforced peace in the north was the loss of most cotton exports for the following five years.

8.2 Agriculture in the Five-Year Plan

On 31 August 1982 the Economic Council approved the first Economic, Social and Cultural Development Plan of the Islamic Republic for the years 1362-66 (1983/4-1987/8). It represented the only effort by the authorities to translate revolutionary Islamic philosophies into a practical programme of action for economic development. Economic priorities established in the plan included (i) educational development, (ii) protection of the interests of the poor, (iii) promotion of economic independence, (iv) provision of a social welfare system, (v) enforcement of minimum standards of food and clothing, (vi) construction of adequate housing, and (vii) pursuit of full employment. Two of the principal pillars that were to sustain the drive towards the plan's objectives were, first, an end to consumerism and, second, the development of domestic agriculture.[18] Hand in hand with the emphasis on agricultural improvement went a policy for reducing the concentration of population and activities in urban areas.

The role of agriculture was to form the 'essential axis of the country's economic development'.[19] To this end an ambitious rate of growth of 7 per cent/year was adopted, supported by a twelve-point policy foreseen as making the target attainable. Among the twelve points were sensible provisions for the integration of programmes in other economic sectors into the agricultural programme and the construction of all those elements of housing, transport, educational and other support services which would permit rapid growth. The plan was remarkably lacking in ideological bias. Traditional and private commercial farming were given equal stress in the allotment of credits and technical services. The plan was very specific in providing for Islamic guidelines to be used in settling the land issue and for the government to refrain from involving itself in land reclamation and management.

Although the planners were optimistic concerning the rate of change for agriculture, they recognized that the likely trend would be for the share of agriculture to decline as a proportion of national income. It was calculated that it would make up 15.3 per cent of Gross National Product in 1983/4 but only 14.2 per cent at the end of the plan period as a result of more rapid growth in both the oil and industrial sectors. The financial targets for agriculture were set as shown in Figure 8.1. Overall, the value added in agricultural activities was expected to rise from 1,540,400 million rials at constant prices to some 2,019,100 million rials by 1987/8. The plan was generous in its treatment of agriculture. Total investment was forecast at

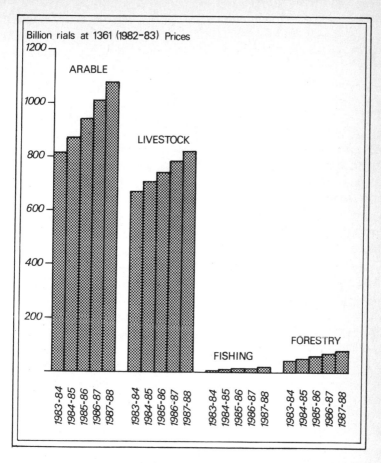

Figure 8.1. Value added in the agricultural sub-sectors 1983/4 — 1987/8.
Source: Central Bank of Iran, *Annual Report,* various years.

2,204,900 million rials over the period, rising from 14 per cent of the total plan investments in 1983/84 to 17 per cent by 1987/88. In respect of details for specific projects it was less precise. Objectives were laid down in a general formula of improvement for all aspects of farming without physical targets being set for an expansion of irrigation or levels of production of particular crops.

Actual performance was inhibited by many constraints. From the beginning the cabinet was less than vigorous in its support for the plan. Although apparently endorsed by the *majles*, the plan was allowed to lapse since the revenues it relied upon were never forthcoming. The year in which it was published was one of comparative financial prosperity as a result of rising oil revenues as volume exports of crude oil increased. However, other factors turned against its implementation. In addition to the lack of political commitment to the terms of the plan, which attracted neither the radicals on account of its politically neutral provisions on matters such as land ownership and nationalization of the means of production nor the conservatives because of its continuing reliance on oil revenues,[20] it also omitted to take full account of wartime conditions. In 1983 the regime was deeply committed to the war against Iraq. A major offensive had begun in July 1982 and there was a conviction among the leadership in Tehran that victory could be achieved given patience and application. Interest in the economy was small. Ayatollah Khomeini and his immediate entourage were naturally inclined against the material concepts on which the plan was based. The very prosperity induced by the inflow of oil revenues in 1983/4 actually militated against the interests of the planners in that there seemed no reason to allocate scarce resources to economic development while the war and basic requirements such as foodstuffs could be provided for out of oil receipts.

Iran's economic position gradually deteriorated after the brief boom period in 1983/4. Financial resources were depleted and pressures grew on foreign-exchange reserves for the war needs. Implementation of the plan, which carried heavy foreign-exchange costs, became increasingly difficult. The crash of the oil price in 1986 and the obstacles to exports created by Iraqi bombing attacks on Iranian oil installations and oil tanker traffic stifled any remaining hopes for a return to the economic development plan. The one real prospect for steady and organized transformation of the economy was lost.

A set of annual budgets for 1983/4-1987/8, which included provision of funds for an economic programme, were substituted for the formal development plan. Allotments of resources to agriculture were generally lower than had been expected by either the ministers responsible for the agriculture portfolio or those who had believed that the revolution would bring a marked change in government policies towards farming.[21] There was little improvement except when political imperatives were at stake. For example, generous credits for farmers were made available just before elections as a means of gaining continuing support from the rural population. Constantly changing budgetary allocations make reliable comparison of initial proposals with final results extremely difficult, though it can be seen from Table 8.1 that spending on agricultural development during the 1980s was generally patchy and small-scale.

In all, there was not the clear determination to put in train major changes in the agricultural investment pattern that might have been expected from a regime that had so strongly stated its backing for the rural community and so roundly attacked previous governments for their neglect. Investment in agriculture was not so easy in a country involved, first, in a phase of post-revolutionary turmoil and, second, in a debilitating foreign war. The flight of many of the agricultural administrators, research specialists, technicians and entrepreneurs during the revolution added to the problems of the Ministry of Agriculture. But the explanation of so manifest a neglect of agriculture in the 1980s, in which there were signs of disinvestment and contraction, cannot lie exclusively in the difficulties of the times. It must be concluded that conviction in the virtues of agricultural development was lacking amongst those who had the power to enforce the allocation of economic resources. The politicians and political groupings appeared to have stronger feelings about what they did not want than positive approaches to the development of farming. A great deal of effort went into stopping land reform or slowing down the return of properties illegally confiscated rather than into the pursuit of measures that would, regardless of ideology, improve the lot of producers in the countryside.

Table 8.1. Spending on Agricultural Development (rials million).

	1977/8	1982/3	1983/4
Distribution of credits by the Agricultural Bank			
Current costs	27,187	28,596	30,420
Irrigation	2,283	13,586	14,170
Orchards	3,437	7,632	9,440
Livestock/Poultry	9,347	42,505	46,595
Rural Buildings	1,489	11,660	16,930
Tractors etc.	—	10,422	18,251
Carpet weaving and handicrafts	—	5,677	5,495
Others	33,414	4,319	4,179
Total	77,157	124,397	145,480
Development Disbursements			
Conservation & Natural Resources	1,950	4,071	10,560
Agribusiness & estates	7,308	5,821	5,572
Improvement of farm produce	2,717	1,204	1,668
Improvement of livestock products	1,786	1,446	1,143
Extension services	1,968	1,596	14,473
Marketing services	7,906	4,120	7
Farm credits	14,020	—	1,100
Co-operatives	3,100	420	690
Agricultural centres, rural and tribal services	—	12,901	12,784
Research	1,322	342	2,038
Provincial credits	—	16,728	16,975
Total	42,077	48,650	67,280

Source: Central Bank of Iran.

8.3 The administration of agriculture and agricultural development

Some of the blame for the slow rate of agricultural development after the revolution lay with the institutional framework which oversaw its management. There was a division of responsibility between the Ministry of Agriculture and the two organizations which in practice operated quite separately from it and each other — the Jihad-e Sazandegi (the Construction Crusade) and the Bonyad-e Mostaz'afin (the Foundation for the Oppressed). There was very little co-operation between these different interests, each with its own separate linkages into the political system of the Islamic republic.[22] Indeed, in 1987 it was proposed by members of the *majles* that the Ministry of Agriculture and the Jihad-e Sazandegi be amalgamated as a means of ending competition between the two.

The role of the Ministry of Agriculture in the years after the revolution was difficult and uncertain. As mentioned above, one of the early policies adopted by the first provisional Minister, Ali Mohammad Izadi, was to recommend that 33,000 members of the ministry's staff be released from their jobs to return to the farms. Understandably, morale took some time to recover from the early naiveté of the revolutionary administration. The situation was made worse by the operations of the Jihad-e Sazandegi, which was fired with revolutionary fervour, supported by many persons in high places and, in its first two years of existence, unencumbered by administrative problems. The Jihad-e Sazandegi took over all development operations first for general rural development and, later, with the permission of the *majles*, for agriculture as a whole. Removal of the ministry's role in the active development of agriculture left it far weaker in preparing and implementing policies for the sector than was previously the case.

Under a succession of ministers policies were rarely consistent though three areas of interest were pursued under all administrations. Land reform was a constant preoccupation. Each minister tried to solve the land ownership problem in his own way. Each sought to re-invigorate the rural co-operative movement and decentralize authority, and each fought to improve the financial resources at the disposal of agriculture.

Involvement with land reform was inescapable in so far as the flight of the landlords, the seizure of lands by the small farmers and labourers, and the disintegration of many of the large-scale farming units run by the state pitchforked the ministry into taking some kind of action to restore order. Unfortunately, the approaches to land reform by the various heads of the ministry were very different. Ministers were never successful in mobilizing option for a single formula to solve the land ownership issue, so that the main prerequisite for renewed growth in the productivity of the countryside — security of land tenure — was not achieved.

The ministry's struggle for an expanding proportion of development funds was far from successful. The claimed level of actual payments was moderately good in 1979/80 at 55,939 million rials but fell dramatically thereafter to recover by 1983/4 to 67,280 million rials.[23] In comparison, industry and mines received almost three times the sums allocated to agriculture in 1983/84. The ministry was unable to get a firm and continuing financial commitment from the government to investment in the sector on a scale that would support a perceptible improvement in production. In real terms the investments made by the state in mechanization, marketing and large agricultural enterprises fell steeply in the period 1979/80-1983/4. This was partly in response to policies that ostensibly favoured the smaller farmers but was also a reflection of the diminishing foreign-exchange resources available for imports. Another area in which the ministry attempted to seize the initiative was in respect of the relative balance between rural and urban interests. Especially in 1984-6 under Dr Abbas Ali Zali, it made efforts to achieve the removal of most food subsidies for consumers and to use the funds thus released as incentives to farmers to produce more goods, but the issue appeared to resolve itself against the rural producers at that stage.

The drive to decentralization was strongly supported by Mohammad Salamati, who took over as Minister of Agriculture from Reza Esfahani in September 1980 and remained until a cabinet change in 1983. His thoroughly laudable objective was to transfer all applied aspects of agricultural administration and servicing from Tehran to the provincial centres. It was proposed to establish some 1,600 ministry-controlled service centres in the main agricultural towns and villages and some 600 centres had actually been set up by the time Mr Salamati left office. His successor maintained the policy of decentralization and very large funds were made available for the programme, accounting for 19 per cent of all investments by the ministry in 1983/4. In that same year 25 per cent of all development funds at the ministry's disposal were allocated to agricultural credits to provincial areas, some through the new centres. It should be noted, however, that, while some of the ministry's activities were moved from Tehran to the provinces, authority for them was not devolved to the local people.

The promotion of rural co-operatives was inhibited by the problems of uncertain land ownership as the legacy of both the shah's land consolidations in farm corporations or production co-operatives and the abortive land reforms of the period after the revolution. Official policies towards the co-operatives were ambivalent in that legislation permitting direct government grants to foster greater activity was not forthcoming.[24] At the same time, different factions within the government took opposing stances on whether the co-operatives should function through local initiatives or be tightly controlled by the state. The position of the Ministry of Agriculture was complicated by its own support for the new agricultural service centres, which it used as channels for ensuring the articulation of its influence in rural areas and which were the agencies through which those credits available to it were distributed.[25] It would seem that preoccupation with the welfare of the service centres left the rural co-operative movement with rather less backing from the ministry than was desirable.

Among the ministry's responsibilities was the supervision of the many organizations that were set up as part of the modernization process or in support of commercial farming units or to manage agricultural commodity transactions. It ran a number of agro-industrial units including the Haft Tappeh and Karun sugar plantations, the Moghan complex and many other smaller enterprises. The central policy towards these units was one of care and maintenance, though one or two, where vigorous management was in evidence, were expanding on the basis of initiatives taken locally.

In the rural areas the ministry operated, as stated earlier, in parallel with two other organizations — the Jihad-e Sazandegi (the Construction Crusade) and the Bonyad-e Mosta'zafin. The former was founded immediately after the overthrow of the shah and in many ways carried the revolution into the countryside. Its principal concerns were the construction of irrigation structures and it took over control of the building of most of the physical developments in the countryside from the Ministry of Agriculture in the early 1980s. In addition, it set up groups to manage cultivation, to 'assist' the farmers with their work and to repair agricultural machinery. Its members organized animal health care and orchard protection through a series of centres scattered throughout the countryside. The Jihad-e Sazandegi attempted to take a major initiative in the land reform and was the agency through which collective farming was brought to the villages under the terms of the Esfahani reform of March 1980 (see section 8.4 below). The apparent rejection of the spirit of the collectivist land reform and the reassertion of private rights in land in 1985/86 left it politically isolated and somewhat diminished in stature. By that time, too, its organizational framework had expanded and it had taken on many characteristics of the central bureaucracy which it had attempted to dislodge.

The Bonyad-e Mosta'zafin (the Foundation for the Oppressed) was heir to the many properties sequestrated from those accused of being associated in a criminal way with the former regime. It took into its ownership considerable areas of land, including agro-industrial units and service companies, thought to amount to some 100,000 hectares. The proceeds of its activities were dedicated to the interests of the poor and deprived, but an element of mismanagement in the mid-1980s left its reputation somewhat diminished. None the less, it managed the largest proportion of the country's commercial dairy industry, producing 20 per cent of all butter and milk. It was very reluctant to give up its rights to estates seized after the revolution even when the tide turned against it and the authorities gradually enforced the return of properties to owners cleared of any guilt by association with the previous regime. The sheer size and spread of its holdings, which included industrial properties, made it a major force in agriculture, but one over which the Ministry of Agriculture appeared to wield negligible influence.

Internal divisions between these various organizations made co-ordinated development of the countryside extremely difficult. The problems were further compounded by a separation of the control of agricultural and water resources. There was an obvious need for carefully co-ordinated action by the Ministry of Agriculture and those who provided irrigation water to farmers. In fact, the allocation of funds to water resources was miserly until 1983/4 and even then was more concerned with the urban needs for potable water, sewage and electricity supplies than the provision of water for irrigation. Difficulties in completing irrigation projects such as the one at Lar were exacerbated by the absence of Iranian specialists and the lack of foreign assistance. It was apparent, however, that policies for water resource development were as confused as those for farming. No clear-cut decision was made on the status of *qanats* vis-à-vis wells and the former continued to decline in importance; no firm approach was given on the question of whether to improve water provision on existing farmland or to reclaim new lands; and construction of major new reservoir structures was not put in hand. The expansion of irrigation through the joint auspices of the Ministries of Agriculture and of Water and Power, without which there could be no reliable and sustainable expansion in agricultural production, was not attempted.

8.4 Land reform, land laws and the effects of insecurity in land ownership

The position of land ownership was far from settled at the time of the revolution. Land reform under the shah's programmes had been more or less continuous since 1963 and there had been considerable release of pressure on agricultural land in the less well endowed areas of the country following the migration to the cities of large numbers of active males during the boom years and their aftermath (1973-8). These changes had not succeeded in removing demands among quite large sections of the farming community for further improvements. In particular, the desire of the peasants to own the land they farmed, first stimulated by the Arsanjani reform of 1962, had never been met since phases two and three of the original reforms of the 1960s had been constructed and implemented in ways that resulted in only small numbers of landless farmers gaining access to plots in their own name and under their own management.[26]

The introduction of agribusiness organizations in areas such as Khuzestan and Dasht-e Moghan added to the numbers of farmers who had lost control of their lands. During the 1970s there was also a period of sustained land purchase by speculators of various kinds. Those in power in the army and government bought agricultural land, some entirely legally. The aggregate effect of their land acquisitions was considerable, however, and in some localities, especially those near to the large cities such as the Varamin plain to the east of Tehran,[27] was to leave large areas in which the local people held the minority holding in land and water resources.[28] The accumulated resentments of the peasants who remained on the land on the issue of ownership were deep and to be found in many areas of the country.[29]

With the revolution and the disruption of law and order throughout much of the countryside the peasantry and others tried to seize land from those who appeared to have lost political support. In addition, there were much larger sequestrations by the many organizations that took part in the revolution. Land was seized from all those condemned for their pre-revolutionary roles, sometimes on trivial or ideological pretexts such as in the case of portions of the 5,448 hectares taken in Gorgan and Gonbad noted above. The matter was further complicated by the withholding of rents from landlords and from the government, the latter in respect of outstanding monies due for lands received during the reforms under the monarchy. In some instances members of farm corporations or labourers on agro-industrial complexes seized land and divided it among themselves. There were unscrupulous individuals who took advantage of the revolution to lay claim to forest lands and uncultivated areas that rightly belonged to the state or were subject to dispute between the state and individual owners in the pastureland areas of the west such as Kermanshahan (Bakhtaran).

The evolution of government policies towards land ownership during the 1980s was slow and tortuous. Practicality and principle were in conflict from the very beginning, while opposition parties exploited the regime's difficulties as far as they were able. The heart of the problem lay in the position of private land ownership in Islamic law. Established law regarding land was not in any way ambiguous — private property was permitted and was to be protected. This single concept was supported by traditional usage ('*urf*) and by all parts of the Civil Code. No grand ayatollah had spoken against private ownership of land or put limits on its extent. The Islamic republic found itself in the unenviable position of having to acknowledge this tenet, whilst wishing for a revolutionary solution to the land question that would clearly offend it. The history of land ownership after 1979 in Iran was one in which the authorities first tried to ignore Islamic law, then attempted to mould it to their own short-term purposes, but were occasionally prepared to accept its conditions in full. Contradictions abounded in such circumstances. Individual plots of land went through a whole gamut of changes in ownership in response to revised interpretations of the law by the government as areas were sequestrated, transferred between state organizations, and allocated to new peasant owners before being repossessed and returned to their original owners.

The practical difficulty arose initially for the government because the revolutionary authorities wanted to keep faith with their followers among the poor and landless by removing their disabilities through land distributed from properties taken out of the hands of the former 'oppressors'. The regime did not wish to be seen at that stage to be protecting the interests of the very class or group against which the revolution had been directed. Sensitivities on this issue were exacerbated by the open dedication of the left-wing guerrilla groups to land distribution. Land reform programmes in the Turkoman areas of the north and in Kurdestan were well advertised and, at a time when the Islamic republic was consolidating its position against the interests of the Mojahedin and Fedayin,[30], it did not wish to appear less than committed to the small farmers and rural poor.

It was equally true that many of the *'ulama* were themselves strongly in support of land reform for social reasons. Local *mollas* in the villages had every reason to give backing to the peasants against the claims of the former landlords, the much hated farm corporations and agri-businesses and the members of the former regime. Islamic law became confused at this stage. Many among the clergy and the government, including the then Minister of Agriculture, Izadi, and his Under Secretary, Reza Esfahani, took the view that considerations of Islam and justice demanded equity of treatment of farmers on the ownership of land and water resources.[31] During the first half of 1979 the Minister appeased the peasantry by accepting the land seizures in rural areas. He went so far as to promise a general land distribution programme in which peasant proprietorship would be fostered.

The first co-ordinated land reform law was published on 1 March 1980. The draft law for the Transfer of Land to Farmers and the Revival of Farmlands was approved by the Revolutionary Council at that time, ratifying the bill of 16 September 1979, which had been a much less radical reform.[32] Rural lands were divided into three categories in the legislation, comprising public lands under the direct management of the government, lands confiscated from members of the former regime, and large farms in private ownership. The first two categories were clearly defined, though only the legal status of state lands was undisputed. The third category, which became known as *band-e jim* (Note C), caused greatest difficulty in definition and was the most offensive in Islamic and political terms. The explanation attached to *band-e jim* indicated that those lands unutilized by large landowners for whatever reasons were liable for distribution to farmers and other qualified persons. The law justified its position by referring to the social and economic circumstances of the Islamic republic. Any structures or installations on lands taken over by the state under the terms of *band-e jim* were to be dealt with at the discretion of the government, though effectively by the appointed committee. Those lands held by large landowners which remained under cultivation were treated in three different groups — first, farms where the owner undertook on-site management of his holding, second, farms run for the benefit of absentee landowners who could demonstrate a need for income from their holdings, and, third, farms managed for absentee landlords who could not establish grounds for receipt of income from their holdings.

Perhaps one of the weakest aspects of the legislation was the definition of the areas to be left to the large landowners. The law proposed that cultivating landlords should receive land equal to three times that area deemed necessary for making an average income. Absentee landlords with a demonstrated need for income from their estates would have a ratio of twice this same notional average holding. These formulas were left to be worked out in practice by the team in charge of land acquisition and distribution. Even worse, the law contained a caveat which permitted the teams to acquire all of the property of individuals at their own discretion and to include government-owned lands in their distribution programme where they saw fit. Such a broad brief within ill-defined terms of reference seemed from the beginning to augur badly for the implementation of the reform, given the considerable opposition that had grown throughout the country against it. At the outset the most positive factor in its favour was its endorsement by three important religious leaders — Ayatollahs Beheshti, Meshkini and Montazeri, who had together formed a committee to supervise preparation of the bill on behalf of the Revolutionary Council.

In order to put the law into effect seven-man teams were set up. They comprised two persons appointed by the Ministry of Agriculture, a representative of the Ministry of the Interior or the appropriate governor-general, a member of the Jihad-e Sazandegi, a representative of the Ministry of Justice or the revolutionary court and two persons to represent the views of the local village council. The legislation was unpopular among many of the senior members of the *ulama* from the very beginning. In *Ettela'at* of 6 March 1980 Ayatollah Sayyed Sadeq Ruhani denounced the bill as illegal. Work under its provisions proceeded despite the resistance of landowners, the merchant class and many supporters of the regime who regarded the law as un-Islamic.[33] The bill was given a one-year validity which was renewed from time to time.

The seven-man teams became closely associated at village level with the Jihad-e Sazandegi. The very enthusiasm of the young members of the Jihad and their predilection for enforced management of distributed lands, some of which they ran on their own account, led to a growing adverse reaction. Attempts to reinstitute the *boneh* system (see Chapter 6) long after its validity had vanished and the perceptible move towards collectivized management of farming were enough to raise a clamour against the reform. Even the peasants themselves were growing weary of the ideological battle in which they were being used as pawns. Farmers almost universally turned against the interference of the Jihad-e Sazandegi where its activities touched land ownership and day-to-day farming operations.[34] The land allocation programme itself did not make rapid and untroubled progress. Husain Mahdavy noted in 1983 that of 1.2 million hectares affected by the land reform legislation at that time only 400,000 hectares had been distributed with a secure title, the balance being held under one-year leases.[35] In practice, the actions of the seven-man committees and their surrogates were perpetuating unrest and uncertainty in rural areas in respect both of ownership of land and other productive resources and of the liabilities of beneficiaries to payment for their land allocations.

The rising tide of discontent led to embarrassments for the government. In particular, opposition to the land reform centred on the fact that the law on which it was based had come into effect before the convening of the first *majles* and had been pronounced as contrary to the laws of Islam by a wide spectrum of grand ayatollahs.[36] On 12 November 1980 the provisions of *band-e jim* were revoked. The turmoil surrounding it was not stilled, however. The Jihad-e Sazandegi and other promoters of the land reform contested the withdrawal of *band-e jim* so that the debate was forced back to the *majles* for a decision.

An attempt was made to modify the law and obtain a ruling on *band-e jim* in late 1982. Proposals made in December were mainly concerned with the rules for payments by beneficiaries and compensation to landowners.[37] Extensive debate in the *majles* resulted in additional provisions, including abolition of the right of the seven-man teams to take over the residual holdings of large landowners once the local formula of three times the average holding had been applied. Orchards were also excluded from their remit, and it was determined that landowners should be fully compensated for assets lost as a result of the acquisitions of the seven-man teams.

The discussion of the land issue was prominent throughout 1983 and 1984 but remained intractable, eventually drawing in the *majles* commission on agriculture and the Council of Guardians of the Constitution as well as the cabinet. In effect, the then cabinet of Mir Hossein Musavi was quite unable to get its land reform measures — largely based on the Esfahani bill of March 1980 — adopted in the teeth of opposition both within the *majles* and particularly in the Council of Guardians. Meanwhile, conditions in the countryside deteriorated as a consequence of the growing level of insecurity of ownership.

The lack of decision-making on the issue left the seven-man teams and their affiliates in a great deal of difficulty since the legal basis of their land allocations was openly suspect. Land grants made on a permanent basis by the teams were not supported by legally acceptable deeds of ownership, while the large areas allocated to farmers on a temporary basis could not be confirmed since their original owners kept up claims for the return of their properties through the courts. The dilemma was epitomized by a speaker at a seminar organized by the seven-man team in Tehran in November 1984, who pointed out that for the three groups of land earmarked for distribution under the original land reform law of 1980, the distribution of category A public lands met with growing resistance, category B lands confiscated from members of the former regime were taken into the possession of the Bonyad-e Mosta'zafin and left undistributed for the most part, while category C, private lands in large farms, were removed from consideration of the distribution programme.[38]

There were constant calls for Ayatollah Khomeini to make a decision on the matter of land ownership. It was felt by all sides that only his considerable authority would be adequate to bring all the parties in the dispute to accept a ruling. Between 1980 and 1985 Ayatollah Khomeini appeared to give a marginal degree of favour to the radicals in his interpretations of Islamic law and the provisions of the constitution. He indicated that the revolution was made by the poor and that their interests should always be paramount. In 1981 a number of requests were made to him to give a decision on the standing of the land reform law, especially in respect of Category C (*band-e jim*) lands. *Ettela'at* reported on 13 May 1981 that the architect of the second of the land reforms, Reza Esfahani, took the stance that Khomeini should either enforce a solution himself or nominate the *majles* to do so. Three days later in the same newspaper it was reported that 101 members of the *majles* had petitioned Khomeini to order reinstatement of the Category C land distribution programme. In the following years there was constant fretting in the *majles*, in the press and in official reports,[39] on the unresolved matter.

Yet the scale of the problem was not overwhelming when measured simply in surface area. A moderately reliable assessment in 1986 suggested that 120,000 hectares of land was taken from the larger landowners under the terms of *band-e jim* and distributed to rural families but without title.[40] At the same time, 750,000 hectares of land affected by legal or other disputes was leased on what were effectively short-term tenancies to 120,000 farming families. Additional areas were supervised by the seven-man teams, including 700,000 hectares of so-called 'waste land' on which 500,000 hectares of farming was carried out by co-operatives, some of them reconstituted *boneh*.[41]

A full evaluation of the situation, first, of the actual distribution of farm holdings before the reform and, second, of the effects of the land reform, was complicated by the usual poverty of statistics. The latest indicator available to the authorities from a full survey was that of the second Census of Agriculture of 1974. Yet clearly much had changed as a result of migration from the rural areas, the other effects of the oil boom of the subsequent years, and the impact of the revolution itself. On the basis of the figures there was a predictable skew towards large numbers of farmers holding a tiny proportion of the total area of cultivated land. The small holders with less than 2 hectares, who were some 43 per cent of owners, held about 4-5 per cent of all land, while the large owners, a mere 0.4 of holdings, took in 15 per cent of all cultivated land. The bias in the distribution was less than in many other countries of the region and oversimplified conclusions drawn at the time were not necessarily helpful.[42] A statistical statement of the situation in Iran for 1972/3, which included a useful separation of dry and irrigated lands, is shown in Figure 8.2.

The importance of the distribution of farms by size categories was that it showed the range of farms that could be affected by the operations of *band-e jim*. It was unlikely that any of the private farms in the size groups of less than 5 hectares would

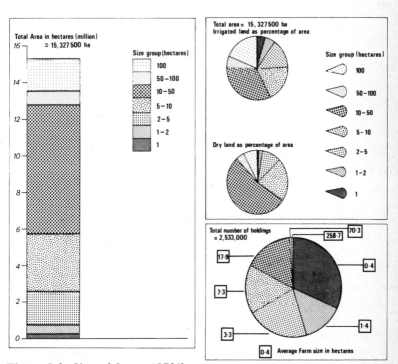

Figure 8.2. Size of farms 1972/3.
Source: Iran Statistical Centre, 1975.

fall under the terms of the land reform, since they would be below the stipulated ceiling of three, later revised to four, times the provincial average. Indeed, it was the opinion of Ahmed Ashraf that most holdings of less than 50 hectares would escape the law on various grounds, including the right of owners to transfer property above the limit to their families. Once account was taken of the lands already designated, including lands returned to their original owners from the former farm corporations and production co-operatives, in the hands of the Bonyad-e Mosta'zafin, allocated to farmers as unconfirmed owners or lessees, and divided amongst the families of the large landlords, Ashraf calculated that approximately one million hectares of land would remain for distribution under the terms of *band-e jim*.[43]

Since the government failed to bring the internal argument on the legality of the reform to a conclusion, this million hectares that remained in the hands of the large landowners was inevitably as insecure as the lands granted without title or guarantee of continuity by the seven-man teams. In effect, at least 2.57 million hectares of land was held by farmers who had good reason to doubt their security of tenure either because they held distributed land on uncertain legal grounds or because they might, as large landowners, believe that the reform together with *band-e jim* would be completed at some future date. A more depressing analysis — though one that was generally accepted in Iran – was that all large holdings over 10 hectares in size on irrigated land were at risk under the terms of *band-e jim*, which would put some 2.5 to 3.0 million more hectares at risk than under the optimistic view outlined above.[44] Thus, some 5.07 million hectares of the country's total of 15 million hectares of agricultural land — more than a third of the total — was put under a greater or lesser cloud as a result of the lack of a clear and binding decision on the status of the land reform.

On 9 June 1986 at the end of *'Ayd-e Fetr* Ayatollah Khomeini called on the government to give the people a role in all economic affairs.[45] This statement did not end the land question but was widely interpreted as discouraging further state intervention in the economy, including agriculture and land ownership. Above all, the views expressed by Khomeini appeared to crush the radicals' final shreds of hope that *band-e jim* could be restored while the Imam remained as the chief arbiter of the Islamic republic.[46] The radicals reserved their positions despite Khomeini's appeals for a compromise between the radical and conservative groups within the government. On 13 June 1986 Hojjat al-Eslam Rafsanjani, speaker of the *majles*, stated in a sermon during Friday prayers in Tehran that 'private enterprise would not be given any more elbow room than it already has', indicating that the battle over the degree of state intervention in the economy continued.[47] In many ways, therefore, the fate of the land reform was still left on the sidelines. Ominously for agriculture, for which stability and a long-term view by the farmers were essential, the ambitions of the radicals were seen as in abeyance rather than abandoned for good.

8.5 The rural community after the revolution

The revolution brought plenty of promises but few measurable improvements to the countryside.[48] In the fields of planning and land reform there were changes but the improvements were either on so limited a scale as to leave the broad area of the countryside untouched or so badly conceived and executed as to leave the villagers worse off than before.[49] Agricultural incomes grew slowly in real terms after the revolution, in contrast to the fall in those elsewhere in the economy.[50] The rate of increase was, however, much lower than had been experienced before 1979 when the rate of growth in value added in agriculture was slightly more rapid, at an average over the period 1973/4-1977/8 of 4.6 per cent, while the rate of expansion of the population was also rather slower.

The alteration in the balance between rural and urban incomes was slight and did very little to alter the fact that rural incomes were markedly lower than those for urban residents. Taking agricultural income alone, it appeared that the income of rural people was some 11 per cent of urban incomes in 1978/9 and 15 per cent in 1983/4. Or to put it another way, it would seem that *total* rural incomes stood at some 59,600 rials per head in 1978/9 and fell infinitesimally to 59,130 rials by 1983/4. Denominated in US dollars, the fall was much larger — from $845 per head in 1978/9 to $657 in 1983/4. Incomes in urban areas were assessed by the present author to have fallen on a similar scale from 108,050 rials per head in 1978/9 to 102,650 in 1983/4.[51] Rural incomes on a *total income* basis represented 55 per cent of those in urban areas in 1978/9 and 58 per cent in 1983/4.

Two points emerge from these calculations. First, agriculture made scarcely any perceptible improvement in its absolute position after the revolution. Average incomes from agriculture alone stayed approximately unchanged mainly because population left the land and pressure on agricultural resources was eased. None the less, measured in US dollar terms, the value of average incomes actually fell from $250 to $230 over the period. Such poor returns to farming activities were a considerable disincentive for a revival in farming by the mass of people. More remarkable is the lowly status of, and returns to, farming, given the high rates of demand for agricultural goods and the relative collapse of formerly rewarding areas of the urban economy such as industry, services and construction. The explanations for this unresponsive performance by agriculture lie *inter alia* in the diversion of domestic demand to consumption of imported goods and the distortion of the domestic market against the interests of domestic agriculture through subsidization of consumers in the urban areas.

Secondly, regardless of the imprecision of the calculations of income outlined above, the magnitudes show that agriculture was a minor component of employment in rural households. The evidence for this was increasingly clear during the 1970s and was confirmed by the surveys of rural households by the Iran Statistical Centre for 1974/5, 1976/7 and 1978/9 (Figure 8.3). Self-employed income from agriculture, together with approximately half the value of household miscellaneous and non-monetary income, accounted on average for *all* rural households for some one-third of total household income according to the surveys.[52] Unfortunately, they did not provide separate data on incomes of land-owning families and landless households, but it was apparent that there were rather diffe-

Agriculture, Revolution and the Rural Community 215

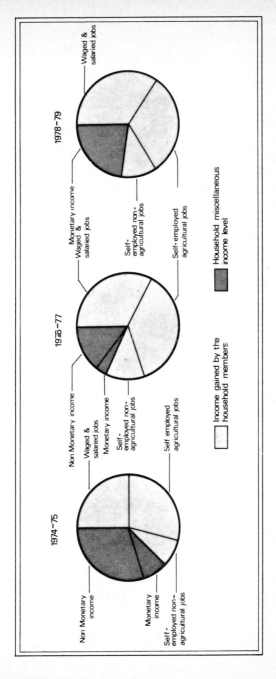

Figure 8.3. Sources of income for rural households.
Source: 'The IRI's rural household survey', *IPD*, 12 August 1986, pp. 10-13.

rent circumstances affecting these two groups with respect to their sources of income and that there were considerable regional variations in the wellbeing of villagers. At the same time, average incomes disguised the polarization between those farmers with small holdings and those with substantial plots of irrigated land. Nevertheless, whatever the sophistication applied to the available data in the analysis, the results remained clear: that there was no movement for change in agriculture or the rural areas towards a rapid growth in personal incomes. The rewards to farming were not improved after 1979. Depending on the statistical basis used, it must be concluded that the wages paid and returns on agricultural investment to the mass of farmers were either poor or very poor.

The proportion of the workforce in agriculture remained steady during the early 1980s at about one-third of the employed workforce, with the total number employed rising from 3.2 million in 1982/3 to 3.3 million in 1984/5. Unemployment and underemployment persisted as problems in agriculture. Of all persons over the age of six years in rural areas in 1984/5 some 37 per cent were actively employed, but once students and housewives were taken from the total, the employment rate rose to 83 per cent.[53] The open unemployment rate, apparently running at 17 per cent, was imprecise in that there was a small element of seasonal unemployment and some abstention from work by those with a private income, but it appeared that unemployment was slightly greater than had been the case in 1976. At that time, according to official statistics[54] the unemployment rate in the countryside was 14 per cent. Underemployment was endemic in many rural areas as a result of the episodic demand for labour in agriculture, the landless labourers in particular suffering prolonged periods without work. There is little evidence that this situation changed after the revolution despite some small attempts by the government to attack the problem by encouraging rural handicraft industries.[55]

Migration from the villages to the towns gave rise to considerable concern after 1979. Scarcely a review of agriculture by the ministries, the *majles*, or the media contained no reference to the drift from the land. The scale of the problem as forecast by the Iranian Statistical Centre was shown only in the records of falling rates of increase in the rural population. The figures available were less than reliable. The Iranian Statistical Centre made a major revision to its estimates of the total size of the rural population (up from 18.5 to 19.2 million persons) and rate of growth (up from 0.9 to 1.9 per cent annually) in 1982/3, the latter correction for reasons that were not entirely clear. Statistical juggling did not disguise the fact that the number of rural residents dropped below that of the urban population in the early 1980s and that the disproportion in the growth rates was large, with urban areas growing more than twice as rapidly as the village populations.[56] In effect, it seemed at the most minimal of estimates that approximately a quarter of a million permanent migrants per year continued to flow into the towns from the countryside. Evidence suggests that this might well be an underestimate of the situation and that as many as half a million a year might be involved in these movements.[57]

The causes of migration were not hard to find. Iranian studies of the problem showed that there was little consensus on the matter. For many, the underlying pressure for farmers to move from the land was connected with insecurity of land tenure. The failure to settle the land reform issue had a depressing effect on the villagers, many of whom had hoped to benefit from a break-up of the larger holdings and the distribution of the so-called 'waste' lands. Unrest, violence and deep uncertainties about the future led to many families moving away to what appeared to be more stable conditions in the towns. This point of view was most strongly put by those on the radical wing of the government who were in favour of completing the land reform, and was supported by the Jihad-e Sazandegi. While it might be agreed that insecurity of tenure was never a good prescription for growth in agricultural production and rural prosperity, it was also possible that the threat of a resumption of the land reform was more of a disincentive for people to remain on the land than disappointment at not getting allotments of land. This was because the implementation of land reform was feared throughout the countryside, while there were few who could be sure of gaining access to new land under the terms of the 1980 law and its amendments. But, whatever the emphasis given within the problem of land ownership and reform, the lack of confidence to invest for the long term because of the insecurity of tenure was a continuing motivation on the part of many village people to seek improved conditions elsewhere.

It was argued, too, that the returns on investment in agriculture were so poor that Iranians were no longer willing to remain involved with cultivation even if they remained in the village.[58] The present author's calculations indicated that there was no distinct segregation in the villages between those who were employed in agriculture and those who were not. Farming made up a greater or lesser element of household income throughout the rural community. What was clear, however, was that the proportion of average rural income from agriculture varied enormously from year to year depending on rainfall, success with the harvest, or the state of the national economy external to the village.[59] As we have seen, the relative position of rural vis-à-vis urban income rose slightly after the revolution, but the absolutely low level of villagers' incomes remained more or less unchanged. And at those standards – at the most optimistic assessment representing about 55 per cent or probably much less[60] of those in the towns — many village residents were unable or unwilling to remain.

The large numbers of small villages and other minor settlements in the provincial areas of the country made provision of modern services such as electricity and water expensive. Even during the height of the oil boom the government stopped short of building public utility links and other facilities in the mass of villages. Events after 1979, including the thoroughgoing upheaval in all sectors of national life as a result of the revolution, the war against Iraq, and the collapse of oil prices in the mid-1980s, militated against a programme of rural renewal. Indeed, very little was done for the urban areas either, though in many cases they had been the object of improvement schemes as part of the developments undertaken during the 1970s. In any event, rural residents were much worse served in every aspect of their living conditions than those living in the towns, according to official figures published in *Ettela'at* in

1981/2. While almost all urban households had electricity and main water supply, the figures were only about 40 and 30 per cent respectively for village households. Only some 2 per cent of rural households had a private car against 18 per cent for urban households; 17 per cent possessed a television set while in the urban areas 73 per cent had a set; 26 per cent of village households had a refrigerator as compared with 87 per cent of urban households. The comparison indicated the qualitative deficiencies of village life and why migrants could expect an immediate and appreciable improvement in their lives through a move to a modern city environment.

Social factors were not without their effect. There had been a powerful rate of internal migration before the revolution, which had meant increasing numbers of rural households with family roots in the towns. Urban employment and an established foothold in the city gave opportunity and in many cases the means for residual family groups still in the villages to move out as part of a continuing process that had little to do with the normal processes of economic and social evolution in this oil-based, Third World economy. The revolution and its aftermath was arguably a contributory factor. Disorder in the economy and periodic breakdowns of the commodities distribution system after the revolution and during the war with Iraq left isolated country areas disadvantaged in matters such as meat supply, access to subsidized foodstuffs and provision of medical supplies. Many rural districts were shaken by political unrest such as occurred in the Turkoman, Kurdish, Baluch and Zagros tribal areas.

The tendency for revolutionary activities to be centralized in Tehran and the establishment of economic and political command systems there inevitably acted as a magnet for those, especially the young, who wanted to be part of the revolution. Even more than under the monarchy, Tehran was made into the political 'central place' where the power, excitement, fervour and action were almost exclusively concentrated, together with the perceived benefits of political change. The villages and small provincial towns seemed at a great distance from the heart of the revolution, which yet was able to affect the lives and standards of living of those in the countryside so deeply.

Increasing economic primacy of the capital and the generally strong assertion of urban political dominance over rural interests after the revolution arose from the displacement of the private sector, including local entrepreneurs, from large parts of trade, management of industry and ownership of land. The seizure of properties and businesses from former supporters of the shah's regime and the takeover of enterprises abandoned by their owners who had fled at the time of the revolution gave the state an immediate extension of its area of economic control. This was augmented as nationalization of the banking, insurance and other financial services was achieved and as trade was taken into the hands of government agencies in the two years following the revolution. Other than in areas which were outside the command of the central government such as Kurdestan, the tendency for rural autonomy to break down in the face of encroachments by government, which was so notable a feature of Iranian life in the 1960s[61] and the 1970s,[62] was greatly increased.

It was little surprise therefore that people from the countryside moved to the towns. The non-monetary benefits of rural living,[63] including freedom from close government scrutiny, absence of rigid controls on economic activities, and the supremacy of the small social grouping in its own village environment,[64] were eroded by the invasion of the villages by government officials, groups of workers from the Jihad-e Sazandegi and the intrusion of revolutionary institutions such as the recruiting agents of the *basij* (volunteer military brigades) which pre-empted traditional systems of social decision-making within the rural family.[65] It could be argued with some merit in these circumstances that the new and expanded impact of urban economic and social dominance on rural communities was an important influence in accelerating the decline in the internal strengths of the village, a process already aggravated by the disintegration of the command and culture system for the *qanats* and the breakdown of the structure of land ownership under the shah's land reform. As a corollary, it might be adduced that even if the revolution of 1979 began without a bias favouring urban as against rural areas,[66] this was not sustained so that by the mid-1980s and possibly earlier the

Islamic republic became concerned principally with urban interests. It drew in soldiers, workers, nomadic pastoralists, farmers, medium-to-large landowners and provincial clerics in great numbers as grist to the urban mill. Predictably, agriculture lost its resilience and neglect of it by the centre was increasingly associated with neglect of farming by the rural community itself.

8.6 Conclusions

The revolution did little positive for Iran's farmers. Commitments by a section of the new regime to agriculture and the farming community, which were undoubtedly seriously intended, were never able to find expression in practice. The revolution proved to be urban-oriented in the sense that those with real power — Ayatollah Khomeini, his immediate associates, the central committees of the revolution and others who controlled the flow of funds within the economy — had neither sufficient enthusiasm for, nor understanding of, agriculture to bring about a drive for the scale of investment of men and resources needed to effect a major change in its fortunes. Indeed, statements in support of agriculture and economic development in general were ultimately cast within an Islamic context as defined by Khomeini and those of the 'ulama who followed his line, which relegated material progress to a low order of priority within the goals of the revolution.

One of the continuing themes of this study has been the importance of relative stability of tenure to farmers. The revolutionary regime was at its most naive in this domain, despite the clear lessons of the adverse effects of the land reforms undertaken by the shah's governments. The aim of the Izadi proposals of 1979 was to regularize the land seizures by peasants and others under the stress of revolutionary conditions, but sought to direct the structure of ownership as quickly as possible into private ownership on the basis of a modified *status quo*. The submersion of these reforms in the much more radical proposals of March 1980 was little short of a disaster. The collectivist ideology behind this legislation was a quaint amalgam of sentiment for the traditional organization of the *boneh*, wrongly thought to be a universal feature of the countryside, together with a harder core ambition among the political radicals for a socialist state-farm structure. The outcome of the collectivist programme of the so-called Esfahani legislation was seen only in scattered geographical areas so that no firm conclusion could be drawn with regard to its intrinsic economic, as opposed to its political, effectiveness. The damage done to security of land tenure arose partly from fears that collective land ownership would spread from the reform areas to other villages.

Much more insidious was the effect of the half-hearted halting of the reform and the elevation of the land into one of the prime areas of political/Islamic contention between the two largest power groups within the regime. In this way, *all* agricultural land fell under a political shadow since the matter could only be finally resolved when it became clear which individual or party voice had command in the *majles* and the Council of Guardians of the Constitution, where the final legislation for the revival or rejection of the land reform would be ratified. A final settlement of the conflict had not been achieved in the mid-1980s and awaited decision, possibly in the post-Khomeini era.

In view of the difficulties over the question of land tenure, where the regime could be accused of indecisiveness but not of lack of interest in farming affairs, it was not surprising that investment in and development of agriculture should lapse into neglect. Expansion of the reclaimed area of land under cultivation was very limited and might well, bearing in mind the scale of rural migration to the towns, have been offset by losses of land through abandonment. The creation of new sources of irrigation water supply, other than those brought into use as projects inherited from the previous regime were completed, was very minor. Mechanization did take place in some villages under the auspices of the ministry and its agencies, but was compensated for by losses of investment by the private sector. In any case it was not pursued at a pace that would have made good the decline in the agricultural labour force or would have reassured the country that something more sophisticated than tractorized cereal monoculture was resulting from the investments made.

In marketing alone the government made some headway. Guaranteed prices for basic agricultural commodities were a boon to farmers, especially where they went hand in hand with short-term credit arrangements. Refurbishment of the grain silos and produce stores was taken up with some vigour by the revolutionary authorities, with excellent results according to official statistics. The marketing system for basic products such as wheat was improved, though there were still complaints on the broader scale of problems with high costs and poor availability of transport vehicles together with some shortages of storage facilities.[67] Re-establishment of the bazaar merchants within the agricultural sector in the mid-1980s, confirmed apparently by Ayatollah Khomeini's 'Ayd-e Fetr speech of 1986, brought the mixed blessings of bazaar responsiveness to the market but also problems of profiteering, hoarding and exploitation of the peasants.[68]

Government policies were influenced very strongly throughout the years after 1980 by the continuance of the Gulf war. Resources were more or less scarce for the entire period 1981-5 despite the brief increase in oil income in 1983/4. For better or worse, the leadership of the country was preoccupied with the conflict against Iraq. It is fair to suggest that the war limited the options that the government might have had to develop the economy, even had it wished to do so. Yet wartime conditions might also have been expected to have created a switch of available financial and material resources to the development of agriculture as a means of reducing the very considerable foreign-exchange burden of food imports within a general attempt to revert to the policies of self-sufficiency espoused in the early flush of the revolution. More remarkably, in a situation of deteriorating oil income and falling reserves of foreign exchange during 1986, no clear move was made to reinvigorate domestic production capacity as a matter of urgency. Neither the short-term problem of keeping the nation fed during the oil price crisis nor the longer-term question of how to enable Iranian farmers to feed a growing number of their compatriots were faced by the government, which acknowledged the need for food self-sufficiency but actually did little to make it possible.

Events in the mid-1980s led to the inescapable conclusion that the government, like most of its predecessors, had no constructive policies towards agriculture that would secure a significant future role for the sector. The environmental constraints on farming, especially in the sphere of water provision for irrigation, were to be — probably unnecessarily — as severe five or ten years after the revolution as they had been before. Disintegration of the traditional water use and farming systems proceeded as fast or possibly faster in the 1980s than at any previous time. Accumulated expertise in managing land and water diminished rapidly as experienced farmers and herders moved away from the countryside. New farming systems to replace the waning traditional sector were notably absent over the greater part of the Iranian agricultural landscape so that output increasingly fell behind population growth and promised to do so for some time into the future.

9 Can Agricultural Self-sufficiency Be Restored?

9.1 Introduction

Iran was able to feed itself and provide a surplus of some agricultural commodities for export until the 1970s. There were exceptional years when drought or pestilence resulted in low output but the overall position was that the value of agricultural exports comfortably exceeded that of imports. The change in Iranian fortunes came suddenly as what appeared to be a very gradual and fluctuating trend towards increasing imports during the 1960s was reinforced in the years 1970/1-1973/4. Thereafter it accelerated rapidly through the subsequent decade. The suggestion that land reform *per se* was responsible in any direct sense for this dependence on food imports, or marked the timing of its initiation, is too simplified a view of the situation.[1] Over these same periods exports of agricultural commodities remained stagnant or declining. As late as 1973/4 agricultural exports more than paid for agricultural imports, which represented approximately 10 per cent of all imported materials. By 1983/4 exports covered a mere 8 per cent of agricultural imports, which accounted for more than 15 per cent of the total import bill. However it might be measured, dependence on food imports and agricultural raw materials had multiplied several fold, while overseas earnings from agriculture had diminished to insignificant proportions.

The dilemma for the government arising from the sudden alteration in the source of the food supply was posed at a number of different levels. Short-term difficulties over funding the foreign-exchange costs of imports became acute very soon after 1977/8 and remained to embarrass the authorities through into the post-revolutionary era. In the same context, governments publicly dedicated to diversification of the domestic economy and ending dependence on oil exports were clearly shown to be failing in their objectives as the main source of exports other than oil declined in importance. The most damaging aspect of the problem was the longer-term economic prospect in so far as the oil exports that provided the funds to finance imports of foodstuffs could not be relied on to last for ever. Indeed, it was forecast in the 1970s that oil revenues could cease to be an expanding and certain form of foreign earnings well before the end of the century.[2] The fear was that increasing resort to food imports would simply create a dependence on overseas suppliers which could not be sustained but which would be eliminated only with the greatest difficulty at some point in the not too distant future.

The choice for government policy lay between permitting an augmenting dependence on imported food and minimizing the decline in domestic output. The import option had the advantage of opening up the Iranian market to cheap overseas supplies that would enable price inflation, endemic in most oil economies, to be kept low to the benefit of the urban population.[3] It might be argued, too, that the country had no room for manoeuvre since rising demand for foodstuffs had to be satisfied at a rate that was altogether beyond Iranian farmers. Imports not only restrained inflation but were also needed to mitigate hardship. Looked at during the heady days of rapid economic growth in the late 1960s and early 1970s, the integration of the domestic into the international economy appeared to carry advantages for an incipient industrializing nation which was expecting to become an important manufacturer and exporter. Within an international division of labour and with an exploitation of comparative advantage, the sacrifice of agricultural self-sufficieny was, it might have been assumed, a small price to pay. As a supplier of crude oil, oil products, petrochemicals and industrial goods, Iran would have been no different from other rich developed nations in importing relatively less valuable agricultural commodities. Such conclusions were reached not only by serving officials and cabinet ministers but also in the 1980s by those altogether antipathetic to the style of economic management before the revolution.[4] The analogy with the developed countries tended to beg the questions of the wasting nature of Iran's oil asset, which made it very different from the mainly renewable industrial resources in use in the economically mature states, and the durability of its own drive to industrialization.

Economic principles and theories are not all, however, as Mahdavy recognized.[5] Political and strategic considerations also bear on the matter. In the years after the first oil boom of 1973/4 the country experienced two sharp lessons in respect of the internationalization of its economy. Among the effects of the windfall financial benefits of the oil price rise was the restructuring of the economy in a way that reduced the relative values of its non-oil components. The traditional agricultural and the modern industrial sectors were all but dwarfed by the growth in oil,[6] and their abilities to operate efficiently were inhibited by the impact of petroleum throughout the economy.[7] The by-product of this, *inter alia,* was to reduce industry's potential to provide alternative exports to oil. Growth of the oil sector pushed up wages, increased other costs of production, and expanded the domestic market for the limited volume of industrial goods that Iran was capable of producing. Dependence on imported foodstuffs in the situation of 1965-73, where an export-oriented industrial economy was in successful course of construction as a long-term substitute for oil, seemed a more reasonable proposition than it did in the mid-1970s, by which time the negative effects of the oil boom — including the chaotic state of industry — were to be seen.

There was some comfort to be gained from the augmenting wealth in foreign exchange arising from the oil boom of 1973/4, which suggested that it would be many years before the country faced a balance-of-payments crisis severe enough to jeopardize food imports. To an extent after 1976 but undisguisedly after 1984, the position changed for the worse so that the government was given a foretaste of the domestic impact of external food dependency in conditions of a declining international market for oil. Despite a variable but generally deteriorating balance-of-payments position induced after 1980 by the war with Iraq and after 1984 by a considerable fall in the volume and value of oil exports, the government found itself unable to cut back radically and permanently on agricultural imports. The lesson was demonstrated that food imports could be lightly undertaken but were less readily cut off. There was evidence also that the problems of stimulating domestic agricultural development were as pernicious in times of attempted withdrawal from buying produce abroad as they had been earlier when they had encouraged the move to external supply. Given the changes in land and water use in the period after 1973/4, the task of reconstituting a healthy agriculture in Iran became, in fact, appreciably more difficult with the passage of time. Not only was there a loss of physical capacity for agricultural production in the countryside, but the scale of demand also grew rapidly under pressure from growth in aggregate incomes and rising population numbers[8] so that local farmers were asked in the mid-1980s to produce twice the volume of goods they had failed to supply earlier.[9]

The political nature of 'dependency' arising from reliance on imported foodstuffs was rarely emphasized by Iranian or other writers during the 1970s. Even Bizhan Jazani in his various tracts on dependent capitalism[10] omitted reference to the problem of food imports, despite a detailed review of the entangling of Iran by the imperialist powers. He was more concerned at the time he wrote about the investments by the USA in Iranian agriculture and industry than the effects on his analysis of reliance on overseas food suppliers.[11] Husain Mahdavy's study of Iran as a *rentier* economy[12] also contrived to ignore this particular problem. Among the stronger warnings to Iranian officials of the perilous nature of their growing addiction to food imports was that of René Dumont of the French Institut National Agronomique de Paris Grignon[13], though Dumont's particular caution came late in the day in view of the much earlier move to large-scale imports of agricultural commodities.

Official policies on the issue were, on the surface at least, mainly in support of the expansion of domestic agriculture and promotion of national self-sufficiency.[14] The First, Second and Third Economic Plans (1949-68) included reassurances that national self-sufficiency would be a government priority. The two subsequent five-year plans (1968-78) failed to follow this lead. In the event, successive governments paid little heed to the stated aims of the plans and opted to make good shortfalls in domestic agricultural production through imported goods when this became necessary during the 1970s and 1980s.

It will be argued in this discussion that the growth of agricultural imports was an inevitable outcome of aggregate government policies towards farming activities and the economy at large. The authorities were to an extent pushed towards imports as an answer to their immediate problems of food supply by their view of the inherent inflexibilities of the traditional agricultural system as well as their failure to promote modern commercial farming. At the same time, the surge in the export earnings of the oil industry during the 1970s at first enabled agricultural imports to be funded with ease as they were elsewhere in the oil-exporting states. In retrospect it is difficult to believe that some increases in food imports were unavoidable in the atmosphere created by sudden wealth in foreign exchange in and after October 1973. Such factors, however, do not entirely mitigate the culpability of those responsible for the economy in the pre-revolutionary period, who permitted the demoralization of the traditional farmers, were less than sensitive in establishing a modern agrarian structure and, as a regime, mismanaged the use of oil revenues during the years of the oil boom. These mistakes accounted for the scale of growth in agricultural imports and the rapid plunge into dependence on overseas suppliers.

The country had, and to a large extent retains, positive advantages for agricultural development. It is proposed here that some components of Iran's agricultural inheritance have been all but lost but that there are important physical and human resources that could permit a steady diminution in the country's imports of agricultural goods to a stage where dependency could be brought within manageable bounds. An analysis is required, however, of the current and likely future balance sheets of domestic production, imports and trends in demand for agricultural commodities to determine whether Iran has the potential to pull back from permanent external dependence on food supplies from abroad.

9.2 An assessment of self-sufficiency

Statistical data in Iran are notorious for their unreliability. This applies as much to present-day materials as those for past years. Government organizations issue quite contradictory information, which for wheat, for example, gives a variation of plus or minus 50 per cent in some years. The Ministry of Agriculture has for many years tended to estimate production on the high side for internal political purposes, while the Iranian Statistical Centre pursued a policy of hard realism with production figures set at much lower levels than those proposed by the ministry. Changing and imprecise methods of classification of imports by the customs administration make definitive quantification of import trends extremely difficult for much of the period under review. Manufactured items such as refined sugar and bulk tea appear to be omitted from the agricultural category so that there is constant under-reporting of the value of imports. What follows, therefore, is in many ways a personal assessment of the position, aided by the base figures drawn from available official studies and the estimates of other observers. Because the statistical base is not reliable sophisticated manipulation of the data would seem misplaced and misleading. The data in Figure 9.1 should be looked at as general indicators for the years in question.

The overall trend in food supply is clear enough despite the variability and patchiness of the data in product or temporal coverage. Iran's ability to feed itself deteriorated consistently for most commodities in the modern period. In 1964 the value of all agricultural imports was approximately $68 million, when sugar was the principal commodity imported followed by wheat. Wheat imports were necessary only in small quantities except in drought years such as 1339 (1960/1), though access to American aid grain proved addictive and drew Iran into use of imported grain. By 1967 the situation was only little changed, though sugar and, to some extent, wheat imports fell. By 1974 there was little apparent deterioration in the position, which was bolstered by a rapid increase in self-sufficiency in sugar as a result of new cane sugar supplies from Khuzestan and an improved output of beet sugar and generally rising wheat har-

Can Agricultural Self-sufficiency be Restored? 231

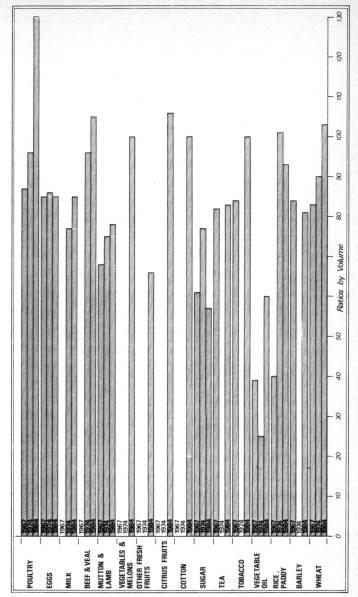

Figure 9.1. Estimates of agricultural product self-sufficiency 1967, 1974 and 1984.
Source: Arezvik, *Agricultural Development, IPD,* 3 September 1985, pp. 6-11; 22 April 1986, pp. 9-11; 23 September 1986, pp. 6-12.

vests. The surge in local demand arising from growing real incomes was reflected in larger imports of livestock, but much of the new demand was either satisfied by relatively good domestic harvests or not met at all. In this situation, food prices rose extremely rapidly at more than 19 per cent during 1974/5 according to the Central Bank, despite the initiation of price subsidies on wheat, meat and sugar.[15]

During the decade following the 1973/4 oil boom Iran became increasingly dependent on overseas suppliers of agricultural goods. Government pricing and subsidy systems aided the trend since they worked against local farmers and facilitated the activities of central public-sector importing agencies and private importers. By 1984 there was not a single major agricultural commodity in which the country was self-sufficient. Dependency ratios ran at between 15 and 20 per cent for most products according to official figures. An alternative estimate of commodity requirements from overseas in 1984 suggested rather higher dependency at 1.2 million tons of wheat, 1.0 million tons of rice, 1.5 million tons of feedgrains, 400,000 tons of vegetable oil and 650,000 tons of wheat.[16] In many ways, it was the sheer range of Iranian agricultural imports by the mid-1980s that made reliance on overseas suppliers expensive in foreign exchange, despite very competitive pricing of foodstuffs on the international market (Figure 9.1). In 1983/4, with purchases of agricultural goods reported by the customs authorities at some $2,796 million, Iran was among the major importing countries of the developing world. This estimate was less than external evaluations of the position. The United States Department of Agriculture and other independent bodies calculated that Iranian agricultural imports were as shown in Figure 9.2. In wheat Iran ranked as the sixth largest importer in the developing countries group for 1983.[17]

Overall, the rise in importance of agricultural imports to represent so large a fraction of the total, in many years standing at more than 15 per cent, signalled Iran's arrival at a state of dependency. Other criteria used included its purchase overseas of more than a quarter of its main food needs. In the mid-1980s wheat and sugar imports were one third of total supplies and red meat more than a quarter.[18] The entrenched

Can Agricultural Self-sufficiency be Restored? 233

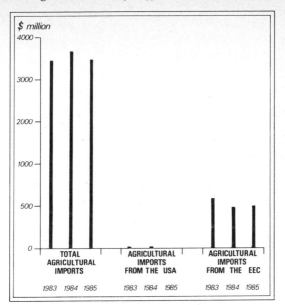

Source: US Department of Agriculture.

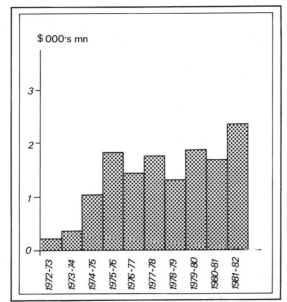

Figure 9.2. Iranian agricultural imports 1983-5.
Source: Central Bank of Iran, *Annual Reports,* various years.

nature of the dependence was also illustrated by the fact that the underlying trend in food imports was upwards and had been so for at least a decade. The Central Bank of Iran reported in 1984 that wheat, sugar and red meat imports rose by 96, 60 and 32 per cent, respectively, over the preceding year.

The origin of agricultural supplies was diverse, but there were some bilateral concentrations such as red meat purchases from Turkey and New Zealand or wheat from Australia and the EEC. During the 1980s a large proportion of purchases of agricultural goods was negotiated on bilateral terms, often on an oil-for-agricultural products basis so that the drain on foreign-exchange reserves was reduced to a minimum. The diversity of sources, the favourable international commercial structure of agricultural trade and the generally low unit prices of food acquisitions abroad removed the political or economic threat from any single foreign supplier.

9.3 The self-sufficiency problem: a balance sheet — the positive factors

It was established in Chapter 2 that the country is not richly endowed with physical resources favouring easy success in agricultural development. This premise remains central to the analysis of any assessment of the problem of self-sufficiency, but there are environmental factors that can be, and in some cases have been for centuries, turned to advantage for cultivation. In effect, in elaborating a balance sheet of elements that work for and against the re-establishment of food self-sufficiency, it will be shown that there are still considerable assets in Iran, which, if exploited in a consistent and well-adjusted manner, could hold out the promise of a reversal of the trend towards accelerating import dependency which emerged during the period from 1974 to the present.

One of the country's main assets is its geographical diversity, though poor communications have to a large extent denied exploitation of this situation in the past. Iran is a country of considerable surface area (some 1,650,000 square kilometres). It contains within its boundaries a remarkable range of climatic and topographic regions. Characteristics of soil, slope, aspect, water supply and drainage combine in all kinds of permutations which, with climatic variations, give the country a wide if not unique regional diversity. The elements producing its great wealth of agricultural regions are in themselves uncomplicated.[19] Iran lies athwart the sub-tropical latitudes which have the attributes of only slightly variable high temperatures throughout the year. These conditions apply in practice only in the southern strip of the country where there is an absence of high relief, which permits the full influence of the latitude to be felt. Elsewhere, sub-tropical temperatures are modified by other regional or local effects such as altitude, continentality, wind conditions and exposure to interactions with adjacent climatic regimes such as the monsoon in the south-east and the rain effects off the Caspian Sea in the north.[20]

The outcome of the very high range of altitude in the Iranian mountain systems, the extreme continentality of the central deserts and the intrusions of air from adjacent zones on the sub-tropical base has been to create a rapidly changing and often clearly demarcated series of principal climatic regions. These in turn are composed of large numbers of sub-regions where more effects of location such as aspect and altitude come into play. The variety of conditions within the principal climatic regions, however they might be defined, is very considerable.[21] This in turn allows a broad spread of plant ecosystems including those that permit a gathering economy of items such as walnuts, hazel nuts and pistachio and the opportunity to cultivate a wide range of field and orchard crops including sub-tropical and temperate varieties without costly artificial support. The possibilities for agriculture opened up by this varied environment were not purchased without cost. High altitudes and extreme influences upon temperature conditions arising from the continentality of central areas of the country made agriculture in most climatic sub-regions subject to often severe thermal and water supply constraints.

An important part of the evolution of Iranian agriculture was the introduction and development of crops and crop varieties that survived and in many cases flourished despite the problems posed by the physical conditions. In western Iran, it was noticed during the 1970s that local varieties of wheat stood up much better even in the short term to the problems of weather and water supply than imported high-yielding varieties, even where agronomic practice with the latter had been as prescribed.[22] Development of varieties of melons and other kurkubits, often with appreciable regional variations, in many ways reached a fine art, matched in few other countries.

Iranian techniques of crop cultivation were unflatteringly dealt with in the literature. Iranian farmers had, apparently, few admirers and most descriptions, whether that of Lord Curzon[23] or later commentators such as OCI[24], emphasized the primitive nature of their practices. Certainly, innovation was not a widespread characteristic of the traditional system on the Iranian Plateau. With the benefit of hindsight and especially in view of the poor record of the modern high technology farms, such as the agribusinesses in the Dez project, it might be suggested that resistance to unproven crops and techniques was, perhaps, largely justified. Indeed, where new crops showed their worth in local conditions, as sugar beet and tea did in the current century and as commercial sunflower seed production did in recent decades, there was little problem in persuading farmers to take up their cultivation on traditional village lands.

The conventional view of Iranian agricultural practices as backward and unenlightened may also be challenged on other grounds. Perceptions of urban Iranians and foreign visitors were influenced by a number of aspects of rural life, three of them glaring to those alien to the system but none of which were material to an evaluation of the efficacy of field practices in farming. First, the agrarian structure, with all its inequalities and oppressive command systems of landlord and peasant, seemed inefficient and outmoded. Secondly, the high manual content of farmwork in the villages looked increasingly primitive in contrast to the mechanized agriculture used in Europe and the agricultural economies of the New World. Thirdly, the landlords, bailiffs and peasants who were active in day-to-day cultivation appeared to be mainly illiterate and ignorant. Yet, in the actual exercise of their farming, many village communities and individuals were profoundly intelligent both in the maintenance of long established techniques and in the use of personal skills.

The range of crafts deployed in the villages was considerable. Some were so obvious that they were taken for granted, such as the levelling of the land for irrigation, in which slopes were handworked to gradients exactly suited to plant requirements, waterflow and drainage[25] without spoiling the soil profile or the fertility of the land. In this respect any comparison with the Dez irrigation would favour traditional farming systems, since at Dez levelling was either ignored and irrigation undertaken on very wasteful lines or was implemented with adverse effects on soil fertility.[26]

Other techniques which illustrate the skill of traditional farming were to be found incorporated in the physical structure of the village lands. The geography of village cultivation that brought together utilization of land and water was simple but extremely effective. Within or immediately adjacent to the residential areas of many villages were irrigated gardens with high walls where a wide variety of fruits and some vines were grown, together on occasions with kitchen garden crops. The walls sheltered the trees and arable crops from the cold, strong winds and grazing animals. Water would be available to the walled gardens as a first priority after household use. Beyond the high walled gardens, especially in settlements around the central deserts and in the east of the country, were somewhat larger fields with lower walls often less than a metre high. Here water would be fed if possible throughout the year even where a *hava bin qanat* provided a seasonally fluctuating water supply. The walls sheltered crops from the wind and helped to reduce water losses through evaporation. Finally, the irrigated cereal ploughlands lay on the downslope below the rest of the village, once again levelled but within broad terraces, depending on the local topography.

The effect of the management of water and terrain for agriculture was to ensure good use of available resources between the different land-use areas of the village domain. It was allied to specialist field techniques in cultivation, including, for example, the deep trenching in vineyards in which slit trenches varying between one and three metres were cut parallel to each other. The individual vines were set in prepared ground at the bottom of the trench which sheltered them from adverse wind conditions and other hazards. Techniques of this kind enabled a considerable variety of crop production.

Terracing was a well developed technique for using water. In the high Zagros and Alborz complex walled terracing was characteristic of farmland, while elsewhere levelled ground was achieved by grading the land along the contour, endowing the cereal ploughlands with a particular sculptured look. Associated with the breaks of slope in terracing were clever sluice and diversion works in the water channels which allowed water to be fed simultaneously along terraces at different heights.

The survival of Iranian farming in less than ideal conditions was managed through diversification of crop production and also through involvement in livestock farming. It was estimated that approximately one third of the value of all farm output came from livestock. Oxen and other draught animals were an integral part of the village system in so far as they provided the motive power for much of the ploughing undertaken. Large numbers of sheep, goats and cows were kept in and around the villages. All but the poorest members of the village community would have some livestock to provide milk, animal fat, wool, fuel and other products vital to the self-sufficient economy in which many of them were involved. Individual holdings were mainly small, comprising two or three sheep and goats, a cow and/or animals for ploughing. Some farmers and larger landlords would keep substantial herds which would be taken out of the village each day to the *dasht* or scrublands in the charge of shepherds. Villages with large livestock holdings also practised various forms of transhumance, in which flocks were moved for many weeks in summer to upland pastures for better pasturage. In the Zagros region some landlords were also affiliated to nomadic and semi-nomadic groups and so were able to take advantage of movements of livestock between the sedentary and the mobile sectors.

Diversification of the rural economy was not confined to farming. Multiple incomes were generated through participation in craft industries such as carpet weaving, hawking products outside the village in the off season, seasonal off-farm labour, and extended migration to a workplace outside the region. Long before the coming of the land reform and the oil boom Iranians developed a taste for cash incomes earned outside farming, though before 1960 the majority had as their objective an ultimate return to their native villages. Most kept their families in their home villages and travelled as unaccompanied workers. Seasonal and short-term migration was, therefore, a support system for much of the rural community until recent years. After 1960 the balance began to change so that most migration out of rural areas became permanent.

The principal craft industry was carpet weaving. Some quarter of a million people were thought to be engaged in carpet weaving in small-scale domestic and commercial workshops in country areas. In Kerman the main organization used was the outwork system under which urban merchants often provided materials to contract weavers to produce carpets and rugs of a design and to a price determined by the merchant.[27] English estimated that approximately 35-50 per cent of people in the larger agricultural/weaving villages around Kerman and some 9-10 per cent in the small villages and hamlets were employed in carpet production in the mid-1960s. The outlying areas of the Kashan carpet industry were structured rather differently according to Mahdavy, with the private weavers acting to a large extent with their own financial resources and on their own behalf despite ultimately channelling their products through the Kashan merchants in either the Kashan or Tehran bazaars.[28] Such patterns were not static. The present author found weaving being introduced on a commercial scale into Ferdaws in the 1960s as a means of creating alternative incomes in a region badly affected by a series of drought years. Weaving by pastoral nomadic groups had regional importance, including the Afshar, the Qashqa'i and the Bakhtiari, among others. Whatever the financial or production arrangements, it will be apparent that the industry was enormously important, often accounting for a half or more of cash incomes in weaving areas.[29]

In the west of the country in those provinces designated as *sardsir* (cool temperate), where three or more months of the winter were spent in enforced idleness from agricultural routines as a result of extremely cold weather, many young males would travel from their villages. In some villages near to Hamadan the author met several families where the men would take trees — maiden fruiting stock and plane tree saplings — grown on their land to sell in distant cities. Other seasonal labour in the city was also undertaken whenever it was available.

The Iranians, of course, had a long history of labour migration for extended periods. Jobs were sought on other farms, in the cities and abroad as a means of raising capital. Fortunate migrants returned comparatively well-off and able to purchase land, water and improved housing. Others were simply enabled to survive their farming and family crises or to pay off their debts. Occasionally one or more members of a family would be permanently absent earning enough to keep the extended family back in the village. Sometimes, members of the family would work within a rotation so that there was a constant source of urban-based income for the rural community. The author met a family from Jandaq in the 1960s who kept going two jobs in a Tehran dry-cleaning/laundry through constantly replacing returnees with new workers from among their number.

Growth of prosperity in the Gulf in the 1950s opened up new and expanded prospects for Iranian labour migrants. It was officially estimated, for example, that there were 18,248 Iranians in Kuwait in 1961, which number increased to 40,842 by 1975.[30] There were probably at least the same number there as illegal immigrants. Most came for the reasons enumerated above rather than as permanent migrants. They worked hard and long hours but often accumulated large sums of money for transfer home, a deal of which found its way into investment in the villages of southern and western Iran.

It can be seen from this discussion that farmers, pastoralists and the rural community at large were skilled at making the best use of limited physical resources. The positive abilities to diversify crop production, maintain a high level of livestock and diversify sources of income enabled the survival of a large rural population. Equally importantly, in the years preceding the land reform the system gave opportunity for a strongly commercial agriculture as well as subsistence; the two ran in parallel within the traditional farming area. Two factors worked towards ensuring the development of commercial agriculture. First, the landlord and peasant system that operated on a large proportion of village lands was managed generally as a share-cropping arrangement.[31] Landlords took their portion in kind, which ensured a creaming off of a surplus each year, part of which was sold on the market by the landlords to give them a cash income. Second, the bazaar merchants, their agents and the informal pedlar sector run by the *pilevar* were integrated into the processes of agricultural production. Urban merchant groups had close connections with both large landowners and other members of the farming population, since the bulk of trade passing through the urban bazaars was in agricultural produce. Merchants relied on exports of agricultural commodities such as cotton, pistachio nuts, dried fruits, hides and tragacanth to fund their imports of manufactured and other goods from abroad. Indeed, the selfsame system was in operation in the 1980s and reinforced through regulations issued in 1984-6 whereby merchants were permitted to retain the foreign-exchange proceeds of exports at favourable rates and to import goods to the value of their exports.

The merchants competed with each other in the market and attempted to increase their domestic supplies of export goods by extending their contacts deep into the countryside. On the one hand, bazaar merchants were substantial money lenders to the larger landlords for agricultural and other purposes. Influence was exerted on other producers through provision of credits, purchases of standing crops, or award of supply contracts. The system was, like most agricultural marketing structures in the developing world, not one that particularly favoured the producers. In some of its operations the bazaar was extortionate, especially where the *pilevars* were concerned. Rates of interest charged to rural borrower-suppliers were often higher than 100 per cent. None the less, in other ways the merchants did play a positive role in moulding agriculture and ensured that a significant volume of produce moved from the countryside to the urban markets.[32]

The decline in the level of state interference in agricultural land holding in the 1980s, symbolized by the failure of the Islamic radicals to force through new land reform legislation, was accompanied by other liberalization of the economy noted in the preceding chapter. This opened the way for a resurgence in farmer-bazaar merchant relations which offered some hope of an increase in the volume and value of agricultural products reaching urban markets in contrast to the depressed performance under state-imposed conventions obtaining in the years 1968-78.

9.4 The self-sufficiency problem: a balance sheet — the negative factors

The virtues of Iranian agriculture were considerable but the damage inflicted on it by economic and political events after 1960 reduced the potential of the traditional sector. Yet reform and modernization policies did little to establish a viable alternative area of production to replace the expanding quantities of agricultural commodities imported in the 1970s. Into the bargain, the many long-standing constraints on improved food output were not mitigated by the application of new technologies. Of the various modernizations of agriculture attempted in the twentieth century, it was that by Reza Shah in the 1930s which had most success and left the rural areas better rather than worse off. In that period steppe lands were reclaimed for use in the northern provinces under new crops such as cotton, tea and mulberry. Government measures to improve agriculture and even Reza Shah's personal land acquisition policies did not diminish the confidence of landlords and other private investors. Self-sufficiency in food supplies was increased as a result of Reza Shah's activities.[33] By contrast, the contemporary period, for all the involvement with land reform, mechanization, introduction of high-yielding crop varieties and supply of chemical fertilisers, saw a slow impoverishment of those traditional attributes that had enabled the survival of agriculture and the maintenance of self-sufficiency in foodstuffs.

9.4.1 *The problem of rising demand*

From the early years of the 1970 the government was faced with the task of feeding a rapidly expanding population that was increasingly prosperous. Demand for foodstuffs grew very rapidly, as is entirely normal during the early phases of rising personal incomes (see Figure 9.3) Income per head rose from approximately 11,375 rials ($175)in the early 1960s to 70,000 rials ($1,000) in 1970 and 141,250 rials ($2,000) in 1980.[34] Whereas Iran was among the low ranking countries of the world with respect to diet as late as 1970, it lifted itself to stand among the best fed of the countries of Asia during the first half of the 1970s.[35] Indicative of the rising standard of diet in the early 1970s, calorie intake per head went up from 2,269 to 2,720 in 1975, and total protein consumption rose from 68.6 to 74 grammes/day, of which animal protein accounted for 15.9 grammes/day in 1972 and 20 grammes/day in 1975. Consumption of high-grade red meat, rice, wheat and sugar rose extremely rapidly. In the case of sugar, by 1980 Iranians stood on a par with Turkey and consumed almost three times as much sugar per head as the Pakistanis and more than five times as much as the Afghans.[36]

Increasing demand for food was linked in the first place with the rural areas. In the wake of the land reform food consumption went up in the villages as farmers, liberated from the rotations enforced by the landlords, diversified their production and took some of their improved living standard in the form of on-farm consumption of their own crops and animal products.[37] In one way, therefore, the land reform increased the tendency for subsistence in those villages it touched. Some of the enforced marketing of agricultural products that had gone on under the landlord and peasant regime disappeared so that, while it is true that overall farm output did not fall as a result of the land reform, the availability of some commercial crops declined.

This pattern changed in the 1970s, when the urban areas experienced incomes rising on average at some 12 per cent/per annum in response to growing government expenditures. Urban demand was augmented by internal migration and by

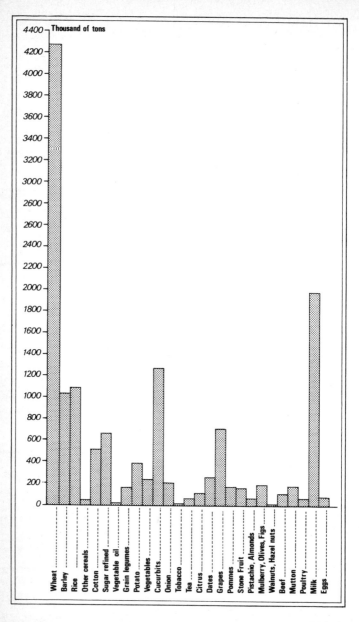

Figure 9.3. (i) Food production 1972.
Source: Booker Agricultural Service, *National Cropping Plan*.

Can Agricultural Self-sufficiency be Restored?

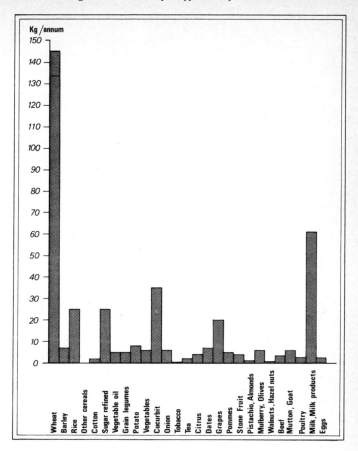

Figure 9.3. (ii) Food consumption 1972.
Source: ibid.

more than a quarter of a million foreign workers and their families brought into the country to support the many new development projects begun in the period after 1973. The foreign element in consumption added to existing trends towards demand for high-quality and more varied foods. Demand in the principal market, greater Tehran, was significantly internationalized as modern shops and supermarkets provided a growing proportion of the retail outlets for foodstuffs. The grading, packing and quality controls that operated in the modern retail sector were beyond the abilities of all but one or two well organized rural co-operatives so that the modern food retail sector found it convenient to turn to foreign suppliers.

Meanwhile, industrialization was proceeding apace, with growth rates averaging well over 10 per cent in real terms throughout the period 1965-78. Food processing played an important role in the expansion of industry, together with manufacture of shoes and other goods that required agricultural raw materials. This group of industries accounted for some 10 per cent, rising in some years to 20 per cent, of all non-oil investment in industry in the period and was supplied to a large extent by imports of raw materials. The sugar refining industry mainly used locally produced cane and beet, though never in a quantity adequate to match the expanding domestic demand. By 1977/8 43.9 per cent, or 510,710 tons, of total supply was provided by imports. The introduction of high technology machinery as part of the modernization of food and agricultural raw material processing meant that a reliable flow of standardized inputs was needed which domestic producers were generally unable to provide. The very high rate of growth in some processing industries would in any case have temporarily outstripped the capacities of most well organized agricultural supply systems. In the case of Iran, the agricultural marketing organization was quite unable to cope with the sudden demands made upon it at a time when other points of demand were also expanding rapidly. Growth in domestic manufacturing industry using home-grown agricultural raw materials had obvious advantages, but also had the effect of reducing the flow of agricultural exports. This exaggerated the net exchange costs of the country's foreign agricultural trade.

Demand for foodstuffs was vitally affected by growth in the size of the population and by its distribution. Before the Second World War the Iranian population was slow growing at some 1-2 per cent annually and was largely located (more than 75 per cent) in rural areas. In the post-war period there was a slow but steady improvement in health conditions as campaigns against malaria, smallpox and other pestilences took effect. Better health facilities gradually spread to most urban areas and, after 1963 and the establishment of the Health Corps, on a modest scale to the villages. The rate of increase in population rose quite rapidly to more than 3 per cent in the 1970s, at which level it persisted through to the mid-1980s. A population which had been a mere 10 million at the turn of the century rose to 20 million by 1955, 40 million by 1980 and 45 million by 1985.[38] This high and sustained rise in the number of mouths to feed set the basic rate at which food output had to increase simply to ensure that the proportion of supply from domestic sources was maintained.

Accelerating migration of rural people to the towns, which saw the ratio of urban to total population rise from 31 per cent in 1955 to 43 per cent in 1970 and 51 per cent in 1980, added to the food supply problem. Fewer and fewer people were able to subsist on their own farm plots. The new urban workers at that time earned a factor of four or five times what they had done in their villages and their propensity to consume foodstuffs once they had arrived in the cities rose accordingly. The very rate at which rural workers left the countryside[39], as was argued earlier in this study, depleted very considerably the pool of skilled labour available for agricultural work and thus reduced the ability of agriculture to respond to the challenge of greater productivity.

9.4.2 *The failure in agricultural output*

The combination of difficulties in promoting improved production after the revolution, noted in the preceding chapter, together with often inappropriate policies towards agriculture and food supply, had a clearly depressing effect on total output. The problem was well defined by the Ministry of Agriculture in a review of the period 1979/80-1984/5 summarized in Figure 9.4, which showed a poor performance in all but a few commodities such as barley, potatoes and onions. Elsewhere, production either just kept abreast of growth in population or fell below it even on the basis of statistics made available by the ministry. Taken at their face value, the official figures for harvests in the period following the revolution suggest declining returns per hectare for some crops, notably wheat, for which the cultivated area rose by 1.7 per cent each year between 1979/80 and 1984/5 but the average growth in output

248 The Neglected Garden

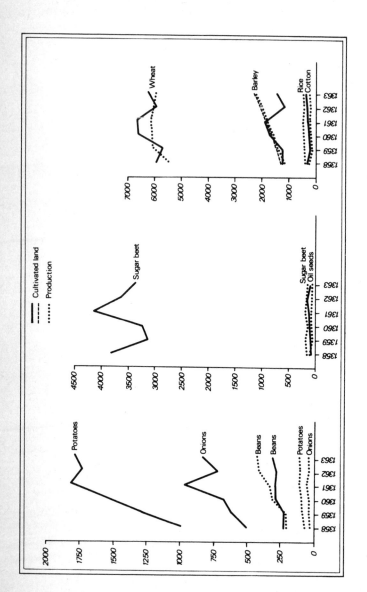

Figure 9.4. Crop production 1358-63 (1978/80-1984/5).
Source: *Yearbook of Agricultural Statistics*, various years.

went up by only 0.4 per cent. The wheat programme was supported by considerable incentives from the ministry so that its low rate of improvement represented something of a failure. The most remarkable increase reported in the five years to 1984/5 was in barley acreage and output, reflecting an option by many farmers to grow an easy crop that required a low labour input and a minimum of irrigation. In like manner, the period also saw a conversion of land from commercial crops such as cotton, sugar beet and sunflower seed to less demanding cereal crops, causing a marked reduction in the value of agricultural output.

Cropping performance was not redeemed by activity in the livestock sector. Growth rates for the principal commodities were all below the rate of increase in the population, except for milk. Red meat at an average of 2.6 per cent, white meat at 2.5 per cent and eggs at minus 3.4 per cent all fell below target. Shortfalls in supply were compensated for by imports or were obviated by rationing curbs on demand. By 1986, however, the government lost the ability to import at will as a result of falling oil revenues. Simultaneously, the national budget came under enormous pressure as government income dropped sharply and the entire food subsidy system became unduly expensive. At that stage the country began to face the stark reality that its policies towards agriculture and food supply were entirely adjusted to the availability of high oil earnings. Rationing of foodstuffs on an increasing scale, either on a coupon basis or through controlled distribution of products in government-managed outlets, was a temporary solution that could hold back rapidly growing demand for an expanding population for a short period and be justified to the population only as a by-product of the war against Iraq.

The balance sheet for agriculture in the mid-1980s appeared to be slipping inexorably into a worsening deficit. The positive virtues of Iranian farming undoubtedly remained but were wasting assets in so far as skills and experience were being eroded by constant losses of rural people to the towns. The area farmed showed some increase but productivity from such land reclamation was low. Land under irrigated agriculture was increased slightly throughout the 1970s and the first half of the

1980s but never on a scale that suggested a major change in Iran's ability to produce crops on a reliable basis regardless of the vicissitudes of the weather. The rate of growth in value added by agriculture, hunting, forestry and fishing for the five years to March 1984 was calculated at 5.2 per cent annually in real terms on the basis of Central Bank of Iran statistics.[40] This rate was historically very high as a result of the effects of two good harvests in 1981/2 and 1982/3 but it conflicts with all other trends in the sector.

9.4.3 Government pricing policies for agricultural goods

The trends in food imports and in the ratio of domestic contributions to total supply over the period since 1970 showed a rise in deliveries from abroad and a decline in the comparative

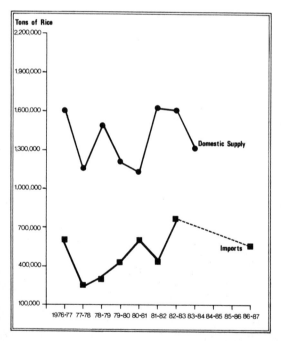

Figure 9.5. Trends in domestic and foreign supplies of rice, sugar and wheat.
Source: 'Present State of Agriculture', *IPD*, 22 April 1986, pp. 9–11.

Can Agricultural Self-sufficiency be Restored?

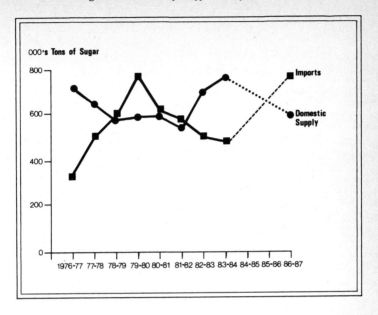

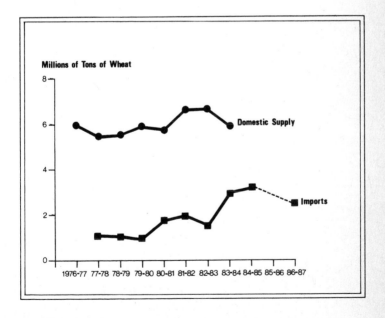

importance of home-grown crops. The situation since the late 1970s is more ambiguous (see Figure 9.5), given the unreliability of the domestic statistics. Quite precipitate swings in the level of imports have taken place, usually accounted for by stock building after a poor year's domestic crop, but occasionally reflecting import allocations that were somewhat eccentric.

Part of the problem, too, arose because of the rationing of several major food commodities. In the case of sugar, rationing was introduced in December 1980 as a means of pegging the rate of increase in imports. The programme was successful, restraining the growth of total consumption and pushing down reliance on imported sugar to less than half of domestic demand. However, the rationing programme resulted not in an increase in home sugar production but in a decline in the per capita consumption from 37.5 kilograms/head in 1979/80 to approximately 30 kg/head in 1983/84.

The rationing programmes went hand in hand with price subsidies on a large range of basic foods so that the pattern of demand and supply was considerably distorted. Even after two good harvests the state spent 108,000 million rials ($1.2 million) on food subsidies through the Organization for the Protection of Consumers and Producers in 1983/4.[41] In effect, the government imported products through state agencies and attempted to distribute them on an equitable basis, absorbing the foreign-exchange costs itself. Such purchases abroad topped up the deficit between domestic output and total home demand to a level officially deemed appropriate. Increasingly in the mid-1980s, the guaranteed price system became a moderately successful tool for guiding Iranian producers to concentrate on basic foodstuffs, though this policy did not appear to be effective in constraining the costs of food imports.

A third factor rose in importance in 1986 as the country became increasingly short of foreign exchange — the need to augment the value of non-oil exports. Exports of agricultural goods fell in real terms throughout the 1970s. The revolution of February 1979 accelerated the decline. The value of agricultural exports in 1978/9 was some $285 million but dropped to $188.3 million by 1982/3, which, taking into account the depreciation of the dollar, was at best a mediocre performance. Cotton sales abroad were negligible in the years immediately following the revolution as a result of political problems that affected the main cotton-growing areas of Gorgan and Dasht-e Gorgan, and the main components of agricultural exports were dried fruits and animal skins, accounting for $125.5 million and $34.6 million respectively in 1983/84.

Subsidies to agricultural producers, including guaranteed prices for wheat and sugar,[42] were extended to some exporters, notably pistachio traders, as incentives. In early 1986 new regulations provided for enhanced treatment of pistachio exporters by allowing high rates of exchange in rials against the US dollar and an ability for exporters to trade in the foreign-exchange allocations they had earned. Unfortunately, foreign markets were increasingly difficult to penetrate for Iranian traders. Some of their former overseas markets for cotton and dried fruits had become very competitive as other exporters had displaced them as a result of the discontinuities in Iranian exports after the revolution or as the new US pistachio growers had joined the market. In conditions of an urgent need for Iran to increase its agricultural exports, the convoluted system of subsidies and production controls became a disadvantage and was used against it by its international competitors.[43]

In fact, the domestic debate on the role of agriculture in the country's new economic circumstances created by the decline in oil income was fraught with political and emotional overtones. Those in favour of an expansion of agriculture used the occasion to repeat their claim that its potential was constantly acknowledged but ignored in practice.[44] In terms of export potential, the only obvious area for advance lay in improvement in cotton sales abroad as cultivation on the cotton-growing lands returned to normal. An expansion in exports of commodities such as dried fruits and pistachio was also possible on a small scale that would help in eking out the country's much diminished foreign-exchange resources. The scope for increasing exports of hand-made carpets and rugs, much of which originated in rural areas and used some domestically produced raw materials, was there to be exploited, given positive policies towards increasing non-oil exports by the authorities.[45] For the bazaar merchants exports of agricultural goods carried the considerable bonus in the mid-1980s of access on a *pro rata* basis to an import quota, which was in itself a great incentive for them to promote farm exports. In all other areas of production, however, any increase in domestic production was needed to make good falls in imports of foodstuffs, as was manifest in government policies which encouraged the cultivation of wheat, rice and sugar growing through the pricing mechanism against the claims of commercial export crops. Expectations of a marked rise in agricultural exports were therefore destined to be disappointed.

9.4.4 *Iran: the oil economy*

The issue of self-sufficiency in foodstuffs came to prominence immediately following the oil boom that began in October 1973. Linkages between Iran as an oil economy and the ability of domestic agriculture to supply the internal demand for food seemed to be close.[46] It might be argued that the influence of oil on the economy before 1973 and certainly before 1970 was perceptible but generally benign, though petroleum as a contributor to national income had continued to increase its importance throughout much of the twentieth century.[47] Without the oil boom of 1973/4 the country could have fed itself for slightly longer — by as much as an extra decade possibly — than in fact it was able to do. Assuming that wheat flour consumption per head had remained constant between 1973 and 1983, it can be calculated that Iran would, *ceteris paribus,* have had a deficit of approximately a quarter of a million tons of wheat grain by the latter year in place of the three million tons of wheat it actually imported.

None the less, a clear discontinuity occurred in the structure of the economy after 1973 (see Figure 9.6). Oil income grew rapidly so that oil and the related service sectors came to dominate the economy. Expenditure of oil revenues was pursued so relentlessly that changes in the rate of growth and alterations in the sectoral composition of national income reflected little of activities elsewhere in the economy. The effects of a rising standard of living on the demand for foodstuffs were

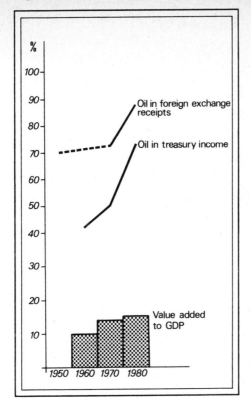

Figure 9.6. (i) The financial and economic role of oil.
Source: Central Bank of Iran, *Annual Reports,* various years.

instant and of a size that gave little hope of a successful or positive response by farmers.[48] During the 1960s domestic food supply was only just keeping abreast of the increase in population, then running at a rate of approximately 2.9 per cent.[49] The trend in population growth, as we noted earlier, was already towards a steep rise and there was no firm evidence that the oil boom accelerated this process. Whatever the causes, the increase in the population intensified during the 1970s to more than 3.0 per cent, which, combined with rising living standards, was more than enough to overstretch the productive capacity of agriculture.

The most damaging aspects of the oil boom for the farming

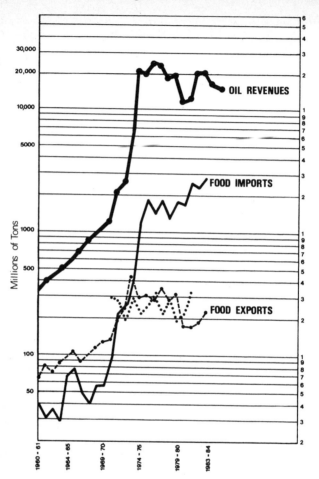

Figure 9.6. (ii) The impact of oil revenues on food imports and exports.
Source: Ibid.

community were its side-effects. Once the decision was taken in 1974 to spend to the full the large oil income that had become available as a result of the oil price rises of 1970-3, the country was opened up to influences that were beyond the government to control. Economic development, in the sense that oil revenues and other financial resources were put to use to establish new productive assets in the non-oil sectors of the economy, was ended. The balance in Iranian imports moved from a bias towards capital goods to the import of consumer items, the latter including goods brought in to support prestige construction projects, military expansion, and strategic plant such as nuclear power stations.[50] The chaos engendered by an undisciplined expenditure of foreign exchange was considerable as congestion grew at the ports, storage facilities were overwhelmed, transportation networks failed to cope and the bureaucratic infrastructure came close to collapse.[51]

The dislocation of the transportation, power and commercial systems was capable of technical solutions, though Iran took more than three years after the start of the oil boom to find them. Even so, there was considerable turmoil within agriculture as a result of these problems. Farmers found themselves competing, for the most part on unequal terms, with the new demands on resources from urban services and consumer sectors.

Movement of farm produce to market was difficult and failures in supplies of fertilisers, other inputs and spare parts for machinery were expensive for a group whose margins were slim. The protracted five-year period of economic disturbance was an additional burden on these same producers. At the same time, the authorities' access to imports made possible by the large-scale availability of foreign exchange encouraged them to answer the growing demand for foodstuffs and agricultural raw materials by overseas purchases. By, in effect, subsidizing food imports, the state automatically prejudiced the interests of Iranian farmers whose products were not subsidized. The pricing policies of the government agencies caused a further adverse trend in farm profitability and reduced the returns to agricultural investment and labour vis-à-vis other areas of economic activity. Marginal lands such as the desert rims and the higher altitude pastures that had formerly been under either sedentary or nomadic use were permanently abandoned as farmers moved to better rewarded occupations.

By far the worst outcome of the increased spending of oil revenues after 1973 was the change in the nature of the Iranian economy. While it might be argued that oil revenues were used fairly effectively in the decade of planned development, 1964-

73, the same was not true after that date. Agriculture in particular suffered badly. Labour was lost to other sectors and above all to urban-based employment. Wage rates took a severe turn for the worse against agriculture.[52] The inflation in all wages rates arising from the oil boom did not pass agriculture by entirely and costs of labour increased even if not as fast as in the urban areas. In consequence, the prices of Iranian produce denominated in the strong rial currency of the period rose against competitors in the international market and, to an extent, in the home market too. Exports became more difficult even where Iranian produce excelled, as in the case of pistachio nuts. Most badly affected was the carpet trade, where labour-intensive operations had traditionally been made possible by employment of low-cost rural labour. Despite the high quality of the Iranian product and an element of non-substitutability because of designs and buyer loyalty, rising prices meant an erosion of the export market. Although carpet exports went up in value by an annual average of 4.8 per cent between 1973/4 and 1977/8 from $108.0 million to $114.5 million, the decline in the US dollar meant that their export performance was poor in real terms.

Rapid material improvements in the urban environment in the 1960s and further developments in the wake of the oil boom of the 1970s were not matched by developments in rural areas — electricity supply, all-weather roads, educational and health centres, together with entertainment facilities, were generally lacking in the mass of small and middle-rank villages.[53] In a very short period of time during the mid-1960s agriculture and/or rural residence became for far too many of the farming community a symbol of absolute deprivation and backwardness in comparison with the larger towns and cities. This perceptible decline in the comparative status of farmers, as the government was ultimately forced to acknowledge in its growth-poles development policies for rural areas, was not simply due to a time lag between benefits becoming available in the towns and then the countryside. Short of great dedication of financial and material resources, which were at the time badly overstretched even in the towns, there seemed little hope that the majority of small rural settlements could be provided with public utilities such as electricity and water in less than a decade or more. In practice, therefore, a more or less permanent cleavage, difficult to quantify in some of its aspects but none the less of great social and psychological moment, had been opened up between traditional farming settlements and the expanding urban centres, to the severe disadvantage of the rural community.

Meanwhile, the economy as a whole became more and more dependent on oil as a result of the events of 1973. Petroleum grew from 18 per cent of national income in 1971/2 to 35.8 per cent in 1977/8. Oil provided 58 per cent of Treasury revenues in 1971/2 but 73 per cent by 1977/8. Foreign exchange derived from the oil sector provided 77 per cent of the total in 1971/2 and 85 per cent in 1977/8. On the other side of the coin, and possibly a better indicator of change, the contributions by the non-oil productive sectors — agriculture and manufacturing industry — diminished rapidly from 30 per cent of national income in 1971/2 to 20 per cent by 1977/8, even using conventional national accounting techniques. The many characteristics of the oil-based economy first observed among the smaller petroleum-exporting states of the Persian Gulf, which Iran had appeared to have avoided in the decade 1964-73, materialized rapidly in the mid-1970s and were never eradicated before the coming of political revolution in 1978/9.[54] For agriculture the negative factors of the changes after 1973 compounded the problems deriving from the ill-advised government policies and neglect of the welfare of villagers of the earlier period. The positive advantages enjoyed by Iranian agriculture noted earlier (section 9.3) had kept a precarious balance between generally severe physical constraints on output and rural survival. After 1973 the balance sheet of factors favouring or impeding agricultural progress was thrust firmly into a deficit from which recovery seemed increasingly unlikely under the economic conditions obtaining between 1973 and 1978. Revolutionary governments debated agriculture with great seriousness but, held back by internal political turmoil in 1979 and embroiled in a war since 1980, actually did little to resuscitate farming.

9.5 Trends in agricultural self-sufficiency: the outlook for the future

In so far as Iran became an oil-dominated economy in the period after 1973, it was predictable that agriculture would decline relative to other areas of the economy and absolutely in the sense of decreasing in significance as an employer, a contributor to real national product or a provider of foodstuffs and other agricultural requirements. Official figures issued during the 1980s were unhelpful. On the surface, agriculture contributed a fluctuating but basically undiminished share of national income (Figure 9.7).

However, oil income itself varied radically from year to

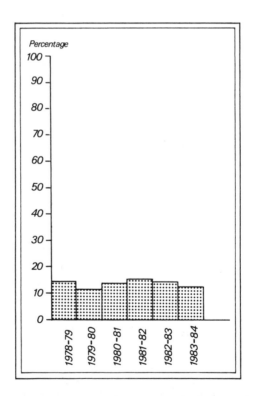

Figure 9.7. Agriculture in GDP 1978/9-1983/4.
Source: Ibid.

year. Wartime conditions inhibited the growth of the oil sector and the volatility of the international oil market added to the erratic nature of the share of oil in national income. The conversion of currencies for purposes of a rial-denominated presentation of national accounts also led to distortion from the use of artificially low rates for foreign currencies. The degree to which the Iranian authorities were likely in the future to be constrained by the nature of the country as an oil-based economy remained in doubt as long as wartime conditions prevailed.

Forecasts of population growth for the medium and long terms showed some diversity. The General Economic, Social and Cultural Development Plan, 1983/4-1988/9, assumed that the population stood at 40,437,000 in 1982/3 and would increase at a rate of 3.1 per cent annually through the five-year plan period to a total of 47,110,000 by March 1989. World Bank estimates suggested that the total population would grow to 53 million by 1990 within a forecast annual growth rate for the 1980-2000 period of 3.0 per cent. Initial reports of the findings of the 1986 census suggest that the rise in population numbers has been high and sustained.[55] A total population of more than 70 million by the year 2000[56] appeared to be modest in view of rising fertility rates. Indeed, past forecasts of population growth have generally turned out to be underestimations.[57] Given the emphasis put on family life and early marriage by the Islamic regime and the negligible advances made since 1978 towards a high consuming modern industrial society with a propensity to limit family size, there appeared to be few natural constraints on population growth. Continuing warfare and poor levels of oil income, together with Islamic views of society, seemed in the mid-1980s to be holding the economy suspended between a peasant tradition and a materialist urban ethic with a consequent bias towards continuing high rates of population growth.

The population was becoming increasingly urban during the 1980s. Official data put the urban resident component at 51.4 per cent in 1983/4 and the five-year plan suggested that it would increase to more than 54 per cent by 1989. But many of the urban dwellers were living in social and to a large extent physical conditions that approximated closely to those in the villages so that the normal expectations of urbanization ultimately bringing down the rate of population increase could not be entertained with confidence. Meanwhile the demographic structure evolved towards one with a heavy concentration in the younger age-groups, whose presence presaged a great increase in fertility and a rising birth rate for the late 1980s and the 1990s.[58]

The scale of the food supply problem was evident in the mid-1980s. By 1986/7 the average supply of domestic wheat per head had dropped by a quarter to 129 kg against 178 kg a decade earlier. Simply to recover lost ground, the country had to increase its output to 9,612,000 tons/year in the five-year period ending 1991/2, an objective altogether outside its capacity on the basis of the experience of the average rate of growth (0.4 per cent/year) in the five years following the revolution. If all imports were to be replaced by domestic supply by the latter year, total home-grown wheat would require to be some 10,433,000 tons on the assumption that there was no increase in consumption per head over that of 1976/7. In effect, an unprecedented expansion in output and area cultivated was called for merely to stand still in a situation of declining imports but increasing population. The position of other main crops such as rice and sugar was perhaps worse than that of wheat.

Ironically it was perhaps fortunate that demand for foodstuffs and other agricultural requirements was inhibited by rationing and a depressed standard of living among the majority of the population. Average incomes per head fell by a third between 1978/9 and 1983/4 according to official sources.[59] On the assumption that a combination of warfare and depressed market conditions for crude oil would continue for an indefinite period, the same depressants on demand that operated in the first half of the 1980s would persist.[60] Equally importantly, expectations among the mass of the population of a rising standard of living were lower in Iran than anywhere else in the oil-exporting states as a result of the Islamic revolution and the sacrifices called for in the war against Iraq. The normal working of demand elasticities would be much modified in such an environment. Frustrated demand might, as in the past, find expression in a burgeoning black market supplied by home producers and importers of smuggled goods in favour of the better-off groups in society and to the detriment of equal distribution of consumption.

Theoretically, pressure on demand, ignoring the operations of the black market, was thus more likely to come through increments in the numbers of mouths to feed rather than through rising consumption per capita. On the most modest assessment of the size of the population in 1985/6 and assuming what appeared to be a reasonable rate of increase of 3.2 per cent a year, it can be calculated that the population should grow by approximately 50 per cent or some 20 million by 1995 and by as many as 35-40 million by 2000. The main choices open to the government appeared to include pushing down demand to 1972/3 levels through constraining supplies, or maintaining diets at 1985/6 levels. The first option would imply a considerable cut-back in imports and an attempt to rely on an improving performance by domestic farmers. A breathing space would be created as dietary standards were allowed to drop progressively to absorb the growing population. In the interim the government would hope for more rapid development of the domestic farming base, and an end to the claims on foreign exchange by the war or an up-turn in the oil market. In order to take up its second option — maintaining food intake and the provision of agricultural raw materials to domestic industry — the authorities would need to achieve (rather than wish for) a perceptible increase in farm production of some 4 per cent annually. Such an improvement would also have to be in basic food commodities and be maintained in the teeth of variable rainfall and other adverse natural conditions. Moreover, increase in output would not have to be purchased by a reduction in the export of agricultural items since their foreign-exchange earnings would be critically important while oil revenues were depressed.

An the analysis in earlier chapters has shown, a dramatic expansion in domestic food supplies appeared to be beyond the immediate capabilities of the country. Failing this, imports would be the only means of augmenting supplies if dietary targets were to be attained. Imports on the scale of the early 1980s at an average of $3,000 million/year could presumably be contemplated only if funds were released as a result of an end to hostilities or if oil income rose appreciably.

If it is accepted that the 1985 Iranian output of food and other agricultural goods cannot be increased by at least 20 per cent by 1990, 50 per cent by 1995 or possibly[61] doubled by the year 2000, then self-sufficiency must be considered out of the question. Reducing demand through rationing and other government controls in the absence of food imports could only hold the line for a short time, given the political problems that would arise during a period of sustained decline in food availability.[62] It is calculated that the drop in dietary levels from 1985 to 1972 standards would be taken up by population growth in less than four years. In effect, under those conditions most conducive to forcing the state into food self-sufficiency — war and shortages of foreign exchange — the sheer scale of the growth of demand in the immediate term from the increase in population numbers alone would make its attainment extraordinarily difficult.

In many ways Iran ceased to be self-sufficient in agricultural trade with the outside world in the early 1970s. In that period the country had a population of 30 million growing at a rate of some 2 per cent/year which was still largely based on the land. At no stage after that time was there a real effort by the government to restore a balance in agricultural trade. Economic changes wrought by the oil booms of 1973 and 1979, demographic movements that have brought on what amounts to a veritable explosion of urban population, and alterations — mainly for the worse — in the country's agricultural resource, enhanced demand for agricultural commodities and also put self-sufficiency in this area beyond immediate reach. The country had become permanently dependent on the outside world to answer its demand for food supplies and other agricultural needs. Short of a thoroughgoing and soundly based expansion of domestic farming sustained over a considerable period of time, the country seemed to have lost for ever the ability to feed itself.

9.6 Conclusion

Iran was not unique in its travails in regard to agricultural policy and rural decline either in the developing world as a whole or among the oil-exporting states. Indeed, its pattern of economic and social change appeared to be destined ominously to conform with those observed in other oil-rich countries so that it might fairly rapidly become in structure (if not in scale) little different from the economies of the smaller oil states of the Gulf, with agriculture relegated to insignificance. This study has attempted to explain some of the reasons why and how this has occurred. Perhaps, too, it will have been clear that Iranian farmers for the most part survived in agriculture until the 1960s despite the marginal economic and environmental nature of their activities. The regret must be that the balance was turned against the farming community with such ferocity. The second and third phases of the land reform during the 1960s reduced the internal strengths of the village communities and their traditional agrarian structures so that they were easy prey for the depredations ensuing from the veritable explosion of urban growth that occurred during the oil boom of the 1970s. In large part the speed and relentlessness of the adverse changes brought about in rural areas were due to the misguided policies of successive governments.

The governments under Mohammad Reza Shah Pahlavi between 1965 and 1978 must take a large share of the responsibility for their inability to understand that the original land reforms would have done better if left to be absorbed gradually by the peasant proprietors.[63] They must also acknowledge their mistakes in unthinkingly exposing the economy to the blasts of the forces of uncontrollable change in 1973-8, which tragically destroyed so much of the valuable gains achieved in economic development in the previous decade. Following the revolution of 1979 many of the errors of the past were repeated — continuing land reform with the uncertainty which it involved, the withholding of adequate investment funds from agriculture, and the reliance on food imports.

The common denominator of official policies towards the rural areas in the present century was neglect of the interests of the farming community. At times this was manifested in a passive way by ignoring villagers' claims in the allocation of funds, material resources and personnel for the development of farming and rural amenities. For the most part, however, neglect expressed itself in the relegation of rural affairs to a role as the servant of urban economic and political welfare. The land reforms were above all tools to rid the shah of his political opponents on the one hand, and to elaborate an urban ideology under the Islamic republic on the other. Pricing policies for foodstuffs were designed to keep down the cost of living in the towns and cities throughout the contemporary period, regardless of the regime in power. The bureaucracy was developed at village level by the central authorities for brief moments in support of locally based and managed development as in 1963-6 and 1979-82. More generally, however, strengthening of the security, ministry and agency networks in the countryside was utilized to impose control from the centre at the expense of local initiative.

The need for the Iranian agricultural environment to be well kept if the investment of centuries in *qanats*, water management, agronomic techniques and farming skills was not to be depleted or even permanently lost was disregarded by governments mesmerized by their access to oil income and deceived by the apparent ease of the expansion of the country's urban industrial base. In subordinating agricultural interests to the benefit of the towns, governments sacrificed one of the few productive economic resources that employed and fed (and could continue to employ and feed) Iranians with a minimum of dependence on the outside world.

The modern period saw an erratic but continuous destruction of the inheritance of the village agricultural landscape and the complex use of elements of water and land which were its main ingredients. Most of the changes, such as the dismantling of the landowning and water management structures, must, as we have seen, be regarded as irreversible. The alternative chosen by the governments was not a different kind of modern farming but a substitution of foreign food imports for Iranian skills in agricultural production. The great Persian garden seemed destined to suffer worsening neglect. With its demise the irrigation systems and village communities, which were for so long the heartland of Iranian culture, will for the most part also disappear.

Notes

Chapter 1 Introduction: Changing Patterns of Agriculture

1. Elwell-Sutton, L.P. *Persian Oil: A Study in Power Politics* (London: Lawrence & Wishart, 1955), pp. 290-1.
2. Greaves, R. 'The reign of Muhammad Riza Shah', in Amirsadeghi, H. *Twentieth Century Iran* (London: Heinemann, 1977), pp. 75-6.
3. Fry, R.N. *Iran* (London: Allen & Unwin, 1954), p. 100.
4. Mahdavy, H. 'Patterns and problems of economic development in rentier states: the case of Iran', in Cook, M.A. (ed.) *Studies in the Economic History of the Middle East* (London: Oxford University Press, 1970), pp. 428-67.
5. McLachlan, K.S., 'Disaster of the oil boom', in *Middle East Annual Review* (Saffron Walden, 1979), pp. 21-8.
6. The role of the USA in encouraging land reform as a means of achieving improvements in the country's political structure is implicit in several interpretations of events that occurred in the early 1960s. Cf. Hooglund, E.J. *Land and Revolution in Iran* (Austin: Texas University Press, 1982) and Greaves, 'The reign of Muhammad Riza Shah', p. 84.
7. Elm, M., 'The need for institutional changes in Iran's agrarian system' unpublished seminar paper, Tehran University/UNESCO, undated. Elm pointed out that 'Iran has more good soil than most countries in the Far East.'
8. Personal communication from Mr Mehran, when Under Secretary at the Ministry of Economy.
9. Central Bank of Iran, *Annual Report and Balance Sheet*, various years.
10. Mehner, H. 'Development and Planning in Iran after World War II', in Lenczowski, G. (ed.) *Iran under the Pahlavis* (Stanford: Hoover Institute, 1978), p. 171. Mehner noted that 'Industrialization, especially heavy industry, received priority in this plan (fourth plan 1968-73), whereas agriculture, water and irrigation projects were reduced'.
11. Hadary, G. 'The agrarian reform problem in Iran', *Middle East Economic Journal* 5/2 (1951).
12. Cf. *The Economist*, 'Survey' 28 August 1976, p. 39, which pointed out that 'The drift of workers from the land into the cities has been taking place at an estimated rate of 8 per cent per year since 1973'.
13. Aresvik, O., *Agricultural Development of Iran* (New York: Praeger, 1976), pp. 235-46.

14. This view was also implicit in the report of Overseas Consultants Inc. (OCI) of the USA, which prepared Iran's first development plan in 1949, *Report on Seven Year Development Plan* (New York: OCI, Vol. III, 1949).
15. At the official level this attitude was represented by the Plan Organization, which, in its 1983/84-1987/88 plan, assumed that the agricultural area could be immediately expanded to give a prospect of self-sufficiency within ten years (see *Economic, Social and Cultural Land Development Plan*, Tehran: Shahrivar 1361, 1982/3, General Policies B, 'Agriculture is the axis of development', p. 11). A similar theme was reflected in the policies of the Ressalat group of the conservative wing of the regime, which called for '...instead of dependence on oil revenue by which farm products are imported for domestic consumption, the revenues should be spent to promote agriculture and encourage the farmers so that the country reaches self-sufficiency in agriculture'. Interview with Ayatollah Azari Qomi in *Ressalat*, quoted in *Iran Press Digest*, (IPD), Tehran, 15 July 1986, p. 6.
16. Curzon, Lord George N. *Persia and the Persian Question* Vol.II, Second Impression (London: Cass, 1966), p. 523.
17. Murray, J. *Modern Iran* (Tehran: Plan Organization, 1950).
18. Hadary, 'The agrarian reform problem.'
19. Lambton, A.K.S. *Landlord and Peasant in Persia* (London: Oxford University Press/Royal Institute of International Affairs, 1953) and *The Persian Land Reform 1960-66* (London: Oxford University Press, 1969).
20. An exception here was Howard Bowen-Jones, who noted in 1968 that 'Thus the illusion of fertility which deceives even many Iranians is created by the little over 6 per cent of utilized land available for perennial and multi-cropping'. Bowen-Jones, H. 'Agriculture', in Fisher, W.B. (ed.) *Cambridge History of Iran*,Vol. I (Cambridge: Cambridge University Press, 1958), p. 576.
21. Langer, F., The Hudson Letter: *Iran: Oil and the Ambitions of a Nation* (Paris: International Herald Tribune, 1975).
22. Katouzian, H. *Political Economy of Modern Iran* (London: Macmillan, 1981), pp. 213-72.
23. International Bank for Reconstruction and Development (IBRD) *World Development Report 1985* (London: Oxford University Press, 1985), p. 212.
24. Economist Publications *Iran, Country Report*, No. 3, 1986, p. 17. '56 per cent of all inputs used in Iranian industry come from abroad.'
25. See Dehbod, A. 'Land ownership and use: conditions in Iran' in *Symposium on Rural Development* (Tehran: CENTO, 1963) p. 74. Other public statements in favour of land reform came from Naser Golesorkhi, Under Secretary of the Ministry of Agriculture in *Echo of Iran*, Tehran, 24 October 1965, and Arfa, H. 'Land Reform in Iran', *Central Asian Journal*, Vol. L, Part II, (April, 1963), pp. 132-7. A contrary view of reform was provided by Naraghi, Y. 'Rural Co-operatives in Iran', unpublished M.A. dissertation, Sussex University, 1978.
26. Development and Resources Corporation *The Unified Development of the Natural Resources of the Khuzestan Region* (New York, 1959).
27. Salmanzadeh, C. *Agricultural Change and Rural Society in Southern Iran* (London: Menas Press, 1980), pp. 21-54.

28. Lambton, *Landlord and Peasant*, p. 343.
29. Beck, L., 'The Qashqai Confederacy', in Tapper, R. (ed.), *The Conflict of Tribe and State in Iran and Afghanistan* (London: Croom Helm, 1983), p. 292.
30. Lambton, *Landlord and Peasant*, p. 288.
31. Apocryphal stories of the 'feudal' absentee landlords, who were ignorant of farming and a crushing burden on their peasants, were in general cirulation before the revolution. Since 1979 such views of the landlords and the system that sustained them have been adopted as the correct interpretation by most official bodies and writers in officially sanctioned publications. A good example of this, available in both Persian and English texts, is the contribution by Ahmad Ashraf, 'Peasants, land and the revolution' in the *Ketab-e Agah on Agrarian and Peasant Issues* (Tehran, Agah, 1982), Ashraf, an informed commentator, took the view that 'Villagers .... tilled and cultivated their lands and a part of the produce was given to the feudal lords, a powerful and wealthy group who lived in the towns. They (the peasantry) worked as the corvee and their youth had to serve the administration as foot soldiers... Although the subject was not tied to the feudal lords' lands and their village, normally they spent their whole life in the villages where they were born.' (The use of the term 'feudal' in relation to Persian landlords is technically incorrect, since their position was not analogous to that of the feudal lords of the manor in medieval Europe. See Lambton, A.K.S., 'Aspects of agricultural organisation and agrarian history in Persia' in Spuler, B. (ed.) *Handbuch der Orientalistik*, Erste Abteilung, Sechster Band, Sechster Abschnitt, Teil 1 (Leiden/Köln: Brill, 1977), pp. 180-1. Professor Lambton suggested that the Muslim conquest destroyed the aristocracy and the feudal system that went with it. See also Lambton, A.K.S., 'Reflections on the iqta' in Makdisi, G. (ed.), *Arabic and Islamic Studies in Honor of Hamilton A.R. Gibb* (Leiden: Brill, 1965), pp. 358-76 and Lambton, A.K.S., 'The evolution of the *iqta*', *Iran, Journal of the British Institute of Persian Studies*, Vol.V (1967), pp. 41-50.
32. Lambton, *Landlord and Peasant*, pp. 259-74.
33. Ibid, p. 395.
34. Pahlavi, Mohammad Reza Shah, *Mission for My Country* (London: Hutchinson, 1961), p. 199, '...it is our agriculture that has the biggest potentialities. And here the *first* big need is land reform.'
35. Afshar, H., 'An assessment of agricultural development policies in Iran', in Afshar, H. *Iran: a Revolution in Turmoil* (London: Macmillan, 1985), p. 69.
36. For Syria, see Manners, I.R. and Sagafi-Nejad, T. 'Agricultural development in Syria' in Beamont, P. and McLachlan, K.S. (eds.) *Agricultural Development in the Middle East* (Chichester: Wiley, 1985), p. 272. The problem in Iraq was reported amongst others by Penrose, E.T. and Penrose, E.F. *Iraq* (London: Benn, 1978), p. 455.
37. There are innumerable references to this aspect of agricultural reform in the current literature circulating in both official and private quarters in Iran. The Minister of Agriculture referred to these self-same problems in a seminar in Tehran in March 1985 as follows: 'The government should have fixed a

minimum price for wheat... so that its cultivation is economical... This would not only encourage the agriculturalist but also relieve the country of dependence on others.' (Quoted in *IPD*, 18 June 1985, p. 7)
38. *Keshavarz*, Tehran, June 1984.
39. The Plan Organization took a different position, in which it calculated that relative agricultural income would decline in the early phases of economic change but would recover later. See *Outline of the Third Plan 1341-46* (Tehran: Esfand, 1344, AHS, 1965-6), pp. 33-5.
40. Nattagh, N. *Agriculture and Regional Development in Iran* (London: Menas Press, 1986), pp. 51-5.
41. Hooglund, *Land and Revolution in Iran*, p. 119.

Chapter 2: The Environment and Agricultural Ecology

1. For a useful basic geography of Iran at mid-century see the British Admiralty, *Handbook: Persia* (London: Naval Intelligence Division, 1945). For general studies of the climate and its effects on agriculture, see Adel, Ahmad Hosain *Abvahava-ye Iran* (Tehran: Tehran University, 1339 AHS, 1960-1) and Momen, Ahmad *Eqtesad-e Keshavarzi* (Tehran: Tehran University, 1341 AHS, 1962-3).
2. British Admiralty, *Handbook*, pp. 34-5.
3. For a detailed account of part of the central desert systems see Mostawfi, A., *Lut-e Zangi Ahmad* (Tehran: Tehran University, 1348 AHS, 1969-70).
4. Honary, M. 'Qanat and Human Ecosystems in Iran', unpublished Ph.D. thesis, Edinburgh University, 1980.
5. Salmanzadeh, *Agricultural Change and Rural Society*.
6. Curzon, *Persia and the Persian Question*, pp. 556 and 559.
7. Oberlander, T.M. 'Hydrography', in Fisher, *Cambridge History of Iran*, pp. 265-6.
8. Ibid. See also Carter, D.B., Thornthwaite, C.W. and Mather, J.R. 'Three water balance maps of south west Asia', *Laboratory Climatology*, xi/1, 1958.
9. Beamont, P., 'The agricultural environment', in Beamont and McLachlan, *Agricultural Development in the Middle East*, pp. 14-20.
10. See the Köppen classification of climates as applied to Iran by Djavadi, C. *Climats de l'Iran*, Monographies de la Météorologie Nationale, no. 54, (Paris, 1966) and Gigon, O. 'Aperçu Hydrographique du Bassin de la Rivière Gamasiab en Iran', unpublished doctoral thesis, Berne University, 1974, pp. 52-3.
11. Gittinger, J.P. *Planning for Agricultural Development: The Iranian Experience* (Washington DC: Center for Development Studies, 1965), p. 73 and Baldwin, G.B. *Planning and Development in Iran* (Baltimore: Johns Hopkins University Press, 1967), pp. 83 and 89.
12. Dorudian, R. 'Modernization of rural economy in Iran', Aspen Persepolis Symposium, conference paper, 1975, p. 12, '...it is foreseen that 2.7 million hectares of improved irrigated land, including 1.4 million hectares of newly irrigated, could be available by 1992'.
13. Plan Organization *Fourth National Development Plan* (Tehran, 1968), p. 145.

14. Mehner, 'Development and Planning', pp. 194-5.
15. The plan was presented in detail in Plan Organization, *Agricultural Frame: Third Plan* (Tehran, 1961).
16. Plan Organization, *Fourth Plan*, pp. 89-115.
17. McLachlan, K.S. 'The Iranian economy 1960-1976' in Amirsadeghi, *Twentieth Century Iran*, pp. 154-5.
18. Bharier, J. *Economic Development of Iran 1900-1970* (London: Oxford University Press, 1971), p. 146.
19. Charles Issawi suggested that the performance was less good. '...in the Fourth Plan only 130,000 hectares were put under irrigation (or one third of target) and 74,000 were improved...' in 'The Iranian economy 1925-1975' in Lenczowski *Iran under the Pahlavis*, p. 147.
20. The calculation was made on the basis of irrigation schemes completed in the period. Hectares (net) for each scheme were aggregated to give the 220,000 hectares total.
21. Central Bank of Iran *Annual Report and Balance Sheet 1349*, pp. 154 and 190. The amounts reported were 21,670 million rials for the Third and 42,006 million rials for the Fourth Plan periods respectively.
22. IBRD *Economic Situation and Prospects of Iran* (Washington DC, 1958), pp. 25-7.
23. Salmanzadeh, *Agricultural Change and Rural Society*, p. 76.
24. Tapper, R.T. *Pasture and Politics* (London: Academic Press, 1979), p. 49.
25. Hadary, G. and Sai, K. *Handbook of Agricultural Statistics* (US Embassy, Tehran, 1949), p. 9.
26. The total area under irrigation of all kinds in 1361 (1982/83) was 3,438,000 hectares according to the Iranian Statistical Centre. See *Salnameh-e Amari*, (Tehran, 1364), p. 314.
27. Price, O.T.W. *Towards a Comprehensive Iranian Agricultural Policy* (Tehran: IBRD, 1975).
28. Planhol, X. de 'Du piedmont Tehrannais à la Caspienne', *Bulletin de l'Association de Géographes Français*, 1959.
29. McLachlan, K.S. and Spooner, B. *A Preliminary Assessment of Potential Resources in Khorasan Province, Iran* (Tehran, Italconsult, 1963).
30. Curzon. *Persia and the Persian Question*, p. 490. 'The Persian peasant... is poor, illiterate, and stolid... he combines a rude skill in turning to account the scanty resources of nature.'
31. Among the estimates are Mahdavi, A.F., 'Agricultural aspects of arid and semi-arid zones in Iran' in *Agricultural Aspects of Arid and Semi-arid Zones* (Tehran: CENTO, 1972), p. 53. More recent data are published annually in *Salnameh-e Amari*.
32. 'Hojjat al-Eslam Fazel Harandi on Land Problems' in IPD, 4 December 1984, p. 17. The discussion reports that the 'Foundation have announced that 100,000 or 120,000 hectares of confiscated land has been allotted by the teams, while this was in fact not true. The amount placed at the disposal of the teams by the Foundation did not exceed 27,000 hectares.'
33. A report in *Ettela'at* of 20 July 1986, 'The role of mechanization in Iran's traditional agriculture' noted that 'Agriculture and the productivity of farm

products per hectare should be improved...'
34. 'Land ownership muddle', *Middle East Markets*, 11/20, London, 1 October 1984, p. 3 suggested that there had been a dramatic fall in livestock numbers. The Iranian Statistical Centre showed the number of sheep falling slightly from 35,952,000 in 1978/79 to 34,605,000 in 1982/83 but all other types of livestock increasing.
35. *Agricultural Census*, First Stage 1352 (Tehran: Iranian Centre, 1354 AHS, 1975-6).
36. 'Agricultural Development' in Survey of World Broadcasts, ME/W1286/A1/6, 8 May 1984.
37. Aresvik, *Agricultural Development of Iran*, p. 70.
38. 'Agriculture', *IPD* 14 May 1985, p. 14. The representative for Babol asked for an explanation of 'the reason for the disorder in the state of agriculture...'.

Chapter 3 National Political Evolution and Agriculture in the First Half of the Twentieth Century

1. Curzon, *Persia and the Persian Question*, p. 490.
2. See Lambton, *Landlord and Peasant*, Chap. VII, pp. 151-71, for a review of the latter half of the nineteenth century.
3. Okazaki, S. 'The great Persian famine of 1870-71' in *SOAS*, XLIX, Part I, 1986, pp. 183-93.
4. For a detailed political study of the period see Burrell, R.M. 'Aspects of the reign of Muzaffar Al-Din Shah of Persia', unbublished Ph.D. thesis, London University, 1978.
5. Lambton, *Landlord and Peasant*, p. 177.
6. Browne, E.G. *The Persian Revolution of 1905-1909* (Cambridge: Cambridge University Press, 1910).
7. Yapp, M.E. '1900-1921: The last years of the Qajar dynasty' in Amirsadeghi, *Twentieth Century Iran*, p. 16. '...in most of Iran life went on much as before...'.
8. Khomeini, Imam *Islam and Revolution* (London: KPI, 1985), p. 256.
9. Millspaugh, A.C. *Americans in Persia* (Washington DC; Brookings Institution, 1946), p. 36.'In practice he (Riza Shah) acted completely contrary to the spirit of the constitution and violated many of its provisions.'
10. Abrahamian, E. *Iran Between Two Revolutions* (Princeton, New Jersey: Princeton University Press, 1982), p. 114.
11. Lambton, *Landlord and Peasant*, p. 181. Professor Lambton summarizes the situation as follows: 'Finally, during the war years the authority of the central government broke down completely. Local leaders, landowners, tribal *khans*, and others were able to assert their virtual independence and to arrogate to themselves again the privileges and immunities which the *tuyuldar* had formerly held.'
12. Garthwaite, G.R. 'The Bakhtiyari Khans, the Government of Iran and the British 1846-1915', *International Journal of Middle Eastern Studies*, 3 (1972), pp. 24-44.
13. Lambton, *Landlord and Peasant*, p. 179.

Notes

14. Millspaugh, A.C. *The American Task in Persia* (London, 1925), pp. 62-63.
15. The Democratic Party proposed *inter alia* state appropriation of large landed estates and redistribution of land to the peasantry during the years 1906-21. See McLachlan, K.S., 'Land Reform' in Fisher, *Cambridge History of Iran*, p. 689.
16. Public domain lands or *khaleseh* were crown lands. The term included state domain and lands owned by the monarch.
17. Lambton, *Persian Land Reform*, p. 35.
18. Bradshaw, D. 'Nomads of Lurestan', lectures given at the School of Oriental and African Studies, Department of Geography, 1979.
19. See Barth, Fredrik *Nomads of South Persia* (London: Allen & Unwin, 1964) for an anthropologist's account of the effects of government intervention in tribal society. For a graphic, though fictional, description of Reza Shah's policies towards the pastoral nomads see Cronin, V. *The Last Migration* (London: Hart-Davis, 1957).
20. Large variations are still to be found in levels of cultivation in the Zagros Mountains. Bradshaw, 'Nomads of Lurestan', reported that cultivation was being forced up above 7,000 feet in 1976/77 as a result of population pressure on limited village land in former nomadic pastoralist territory.
21. In November 1939 a law was passed to improve traditional share-cropping arrangements, a clear recognition by the government of the inadequacies of the system. The law was not put into practice.
22. Lambton, *Persian Land Reform*, pp. 35-6.
23. Savory, R.M. 'Social development in Iran during the Pahlavi era' in Lenzcowski, *Iran under the Pahlavis*, p. 91.
24. Banani, A. *The Modernization of Iran 1921-41* (Stanford University Press, 1961), p. 73.
25. Baldwin, *Planning and Development*, p. 12. Baldwin concluded with respect to Reza Shah's industrial establishments that 'A few of these plants did reasonably well, but many began to decay as soon as they had been finished because of the shortage of technical personnel, managerial inexperience, poor location, and wartime problems with raw materials and replacement parts.'
26. Curzon, *Persia and the Persian Question*, pp. 627-8.
27. A summary of the history of the construction of telegraphic communication between Europe and Asia through Persia is given in Wilson, A.T. *The Persian Gulf* (London: Allen & Unwin, 1928, Third Impression, 1959), pp. 266-9.
28. Baldwin, *Planning and Development*, p. 11.
29. Bharier, J. *Economic Development*, pp. 203-7.
30. Mahdavy, 'Patterns and problems'.
31. Williamson, J.W. *In a Persian Oilfield* (London: Benn, 1927), pp. 169-89.
32. Bharier, *Economic Development*, p. 159.
33. Banani, *The Modernization of Iran*, dedicated his entire book to the achievements of Reza Shah, and presents an interesting case, not accepted by all students of modern Iran, for regarding the monarch as a reformer and

modernizer.
34. Fry, R.N. *Iran* (London: Allen & Unwin, 1954), p. 80.
35. Wilber, D.N. *Riza Shah Pahlavi, the Resurrection and Reconstruction of Iran 1878-1944* (New York: Exposition Press, 1975), p. 17.
36. Baldwin, *Planning and Development* notes (p. 11) 'It is not surprising that throughout Reza's regime transport received the largest share of resources.' By 1938 18 per cent of the general budget (excluding Ministry of War) was allocated to transport and communications. In the same year agriculture received only 4 per cent.
37. Wright, Sir D. *The English Among the Persians* (London: Heinemann, 1977), p. 179.
38. Cottam, R.W. *Nationalism in Iran* (Pittsburgh: Pittsburgh University Press, 1964), p. 196.
39. With respect to the role of the USSR see ibid. Greaves, 'Reign of Muhammad Riza Shah', p. 54 outlined the British position on profiteering by large landowners and the ineffectual efforts of the Iranian Government to bring the matter under control.
40. Bharier, *Economic Development*, pp. 48-9.
41. A graphic account of the problems of shortages as seen from an Iranian point of view was given by Katouzian, *Political Economy of Modern Iran*, p. 143, who described the food supply situation as 'a great scarcity of goods and... hunger and famine in towns.'
42. Lambton, *Landlord and Peasant*, pp. 242-4.
43. McLachlan, 'Land reform in Iran' in Fisher, *Cambridge History*, p. 687.
44. Okazaki, S. *The Development of Large-Scale Farming in Iran* (Tokyo: Institute of Asian Economic Affairs, 1968), pp. 14-21.
45. Lambton, *Landlord and Peasant*, pp. 244-53.

Chapter 4 The Context for Change: Agricultural Development and the Move towards Land Reform 1946-63

1. Katouzian, H. 'Land reform in Iran: A case study in social engineering', *Journal of Peasant Studies* (January, 1974), pp. 220-39.
2. Lambton, *Persian Land Reform*, p. 40.
3. Professor Lambton made the point that the Social Security Law of 1953 placed powers of adjudication in landlord-peasant disputes over the interpretation of the Full Powers Act with the bureaucracy, effectively ensuring that conservative interests would prevail. ibid.
4. Elm, 'Need for institutional changes', p. 10.
5. Hadary and Sai, *Handbook*, p. 19.
6. Kamaly, M.B. 'Iran's experience in agricultural planning', Seminar on Agricultural Planning, Tehran, CENTO, 1971, p. 25.
7. Amuzegar, J. *Technical Assistance in Theory and Practice* (New York: Praeger, 1966), pp. 78-9.
8. Lambton, *Persian Land Reform*, p. 52.
9. Ibid., p. 54.
10. Ghahraman, F. 'Water Rights in Iran' unpublished seminar paper, University of Tehran/UNESCO, undated.

11. Bharier, *Economic Development*, p. 49.
12. Ibid., p. 90.
13. Broekel, R.M. 'Land reform in Asia', *Research Reports* (Washington DC: 1951), p. 95.
14. See Lambton, A.K.S. 'Khalisa' in *Encyclopaedia of Islam*, new edition, p. 979, for a full history of the law relating to crown lands.
15. Pahlavi, Mohammad Reza Shah *The White Revolution of Iran* (Tehran: 1967), p. 33.
16. Amuzegar, *Technical Assistance*.
17. Credits from abroad amounted to $357.6 million for the Plan Organization, of which $317.8m. was actually drawn upon.
18. Hadary, 'The agrarian reform problem', p. 1.
19. Okazaki, S., 'Agricultural mechanisation in Iran' in Beamont and McLachlan *Agricultural Development in the Middle East*, p. 186, Note 2.
20. Hooglund, *Land and Revolution*, p. 45.
21. For a useful summing up of the scale and utility of the foreign aid programme see Baldwin, *Planning and Development*, pp. 200-4.
22. Halliday, F. *Iran: Dictatorship and Development* (Harmondsworth: Pelican, 1979), pp. 256-7.
23. Bharier, *Economic Development*, p. 95.
24. Katouzian, *Political Economy*, p. 225.
25. The second plan period ran from 2 September 1955 to 2 September 1962.
26. Zarnegar, K. (ed.) *The Revolutionizing of Iran* (Tehran: International Communicators, 1973).
27. See, for example, 'Landlords vanishing from Iran', *The Times*, London, 1 October 1964, p. 13. Also Denman, D.R. *The King's Vista* (Berkhamsted, Geographia, 1973).
28. The association was based in Tehran but was never given a chance to emerge as a real force since its leaders were subjected to intimidation and the landlord class as a whole made extremely unpopular through a government-supported propaganda campaign that exaggerated the misdoings of the landlords. In English see *Iran Almanak 1963* (Tehran, Echo of Iran), pp. 398-400.
29. Large areas of land endowed for public religious purposes gave the shrines, religious schools, theological colleges and other charitable Islamic institutions an independent income. Some 2 per cent of cultivated lands were in *vaqf-e 'amm*, see McLachlan, 'Land reform', p. 687.
30. *Iran Almanak 1963*, pp. 391-2. An oblique and fascinating view of the motivations within the *'ulama* during 1963 is given in Mottahedeh, R. *The Mantle of the Prophet: Religion and Politics in Iran* (London: Chatto and Windus, 1986) especially pp. 242-7 and 308-10.
31. Beck, L., 'The Qashqai Confederacy' in Tapper, *The Conflict of Tribe and State,* pp. 301-2.
32. Lambton, A.K.S., 'A reconsideration of the position of the marja' al-taqlid and the religious institutions, *Studia Islamica*, Paris (1964), p. 120.
33. Harney, D. 'Review article', *Asian Affairs*, Vol. XVI, Part I, February 1985, p. 71.
34. Disputes concerning the death toll left the matter unresolved. See Zonis,

M., *The Political Elite of Iran* (Princeton, N.J.: Princeton University Press, 1971), p. 63 and Katouzian, *Political Economy*, p. 228.
35. The continuous influence of Khomeini cannot be ignored. See for example Abrahamian, *Iran Between Two Revolutions*, p. 445.
36. Ibid., p. 426.
37. Cottam, *Nationalism in Iran*, p. 306.
38. Ibid., pp. 306-7.
39. Writing on the reform in 1964, Fredy Bémont was moved to observe of its inauguration in Maragheh 'Ce fut le début d'une tournée glorieuse du souverain à travers le pays', Bémont, F. *L'Iran Devant Le Progrès* (Paris: Presses Universitaires de France, 1964), p. 45.
40. Jacobs, N. *The Sociology of Development: Iran as a Case Study* (New York: Praeger, 1966), pp. 53-62. '...the villages were not able to bond together to challenge the center's political prerogatives' (p. 60).
41. The spread of telecommunications through the new microwave system set up with technical and financial aid from CENTO in the 1960s was important in this respect. The *bi sim* enabled the centre to communicate rapidly with the rural periphery.
42. Naghizadeh, M. *The Role of Farmers' Self-Determination, Collective Action and Cooperatives in Agricultural Development* (Tokyo: Institute for the Study of Languages and Cultures of Asia and Africa, 1984), p. 304.
43. Naraghi, 'Rural Co-operatives in Iran', p. 27.
44. Lambton, *Persian Land Reform*, p. 215. 'The Additional Articles ...could hardly fail to weaken the spirit of the original reform.'
45. Katouzian, *Political Economy*, pp. 238-9.
46. Cf. Rassekh, S., 'Planning for change' in Yar-Shater, E. (ed.) *Iran faces the Seventies*. (New York: Praeger, 1971), p. 161.
47. See the description of problems affecting agriculture in Economist Intelligence Unit, *Iran*, No. 27 (1962) p. 10. The writer reported the occurrence of earthquakes, drought and plagues of locusts in that period.
48. Many economic development projects were adopted in 1964 but only showed in investment and/or physical forms later. See McLachlan, 'The Iranian Economy', pp. 141-7.
49. An unkind view of this group was given by Katouzian, *Political Economy*, p. 234. A 'band of intellectual and technocratic mercenaries whom the shah had been nurturing in the Progressive Centre Club — individuals who would be ready to sell their services, sometimes even their souls, to the founts of power, regardless of their constitution, ideology, economic strategy, political attitude...'
50. Hushang Ansari, who started as Minister of Information in the cabinet reshuffle of summer 1966, and became Minister of Economy in the second Hovayda cabinet, set up an administration of comparatively young but experienced and intelligent middle-ranking officials, permanent secretaries and assistant ministers. The group, including, among others, Hasan Ali Mehran and Dr Najmabadi, proved efficient and creative in managing economic policy. They played an important role in the controlled and far-reaching economic expansion of the years preceding the oil boom of 1973/4.

Chapter 5 Water Resources and Irrigation Cultures

1. Hadary, and Sai, *Handbook*.
2. Ibid., p. 83.
3. Booker Agricultural Services Ltd. and Hunting Technical Services Ltd., *National Cropping Plan*, Interim Report (Tehran, 1975).
4. Murray, *Modern Iran*, pp. 72-3.
5. Lambton, 'Ma"' in *Encyclopaedia of Islam*, pp. 870-2.
6. Ghahraman, *Water Rights*, p. 4, claimed that field crops should be covered by water, tree roots should be covered, and date palms should be inundated as far as the main tree trunk.
7. Ibid., p. 5.
8. Lambton, A.K.S. 'Agriculture in medieval Persia' in Udovitch, A. (ed.) *The Islamic Middle East* (Princeton NJ: Princeton University Press, 1981), pp. 283-8.
9. Ibid., p. 287.
10. Bonine, M., 'From *qanat* to *kort*: traditional irrigation terminology', *Iran*, Journal of the British Institute of Persian Studies, Vol. XX, (1982), pp. 145-59. Michael Bonine examined the continuity of traditional practice and terminology in irrigation in the area of Yazd Province with special reference to *qanats*. His article is amongst the best of sources on the local administration of *qanats*. A clear lesson to be drawn from Professor Bonine's work is the variability of practice and nomenclature from region to region as expressions of the vitality of *'urf* in the countryside.
11. Plan Organization, *Review of the Second Seven Year Plan Program* (Tehran, 1960), p. 31.
12. Ghahraman, *Water Rights*, p. 24, recommended that no new investment be undertaken in *qanats* and that a law be passed discouraging their construction.
13. Aresvik, *Agricultural Development*, pp. 65-6, showed enthusiasm for greater use of water pumped from wells.
14. For example, those laid down by al-Mawardi cited by Lambton, 'Ma', pp. 870-1.
15. Cf. the caveats registered in this regard by OCI, *Report*, p. 130.
16. Bonine, 'From *qanat* to *kort*', p. 145.
17. Mahdavi, M. and Anderson, E.W. 'Water supply system in the margin of the Dasht-e-Kawir (Central Iran)', *Bulletin* of British Society for Middle Eastern Studies, 10/2 (1983), pp. 135-6.
18. Ibid., p. 140.
19. Oberlander, 'Hydrography', pp. 269-70.
20. Ibid., p. 268.
21. This represented adequate water to keep twenty-five plough teams in operation on irrigated land. The nearby Deh-e Juimand *qanat* yielded only eight *juft-e gav* equivalent.
22. Local farmers in Ferdaws claimed that normal yields from *qanat* and well-sourced irrigated fields were 50-fold, while dryland grains could reveal as much as 100-fold.
23. Bradshaw, D., various research materials dealing with Lurestan presented as seminars at the School of Oriental and African Studies, London. A

more complex set of economic relations in which a symbiosis of pastoralist and peasant operated was described in Lurestan by Jacob Black-Michaud in *Sheep and Land* (Cambridge: Cambridge University Press, 1986). The struggle between groups over land is central to this analysis, though it is cast in the context of an emergent capitalist structure.
24. Barth, *Nomads of South Persia*, pp. 103-11.
25. Lambton, 'Ma', p. 866.
26. British Admiralty, *Handbook*, p. 431.
27. McLachlan, K.S. Iran, *Encyclopaedia of Islam*, V, pp. 6-7.
28. Crow, N. et al *Iran: Studies in the Agriculture of the Gorgan Plain* (London: School of Oriental and African Studies, 1979), p. 11.
29. Vertot, M. *La Riziculture dans la Région Caspienne* (Tehran: Cotha-Sogreah, 1960), pp. 4-5.
30. Estakhr, see Bonine 'From *qanat* to *kort*', pp. 154-5.
31. The *karez* of western Afghanistan, the *falaj* of the Arabian Peninsula and the *foggera* of North Africa show similarities to the *qanat* as indeed do the *manbo*. See Okazaki, S., 'Qanat and manbo', *Chi'li*, (Geography), 30/6 (June 1985), pp. 76-84.
32. Okazaki, S., 'Agricultural mechanization in Iran' in Beamont and McLachlan, *Agricultural Development in the Middle East*, pp. 176-9.
33. Okazaki, *Development of Large-Scale Farming*, p. 51.
34. Okazaki, *Agricultural mechanization*, pp. 172-4.
35. Aresvik, *Agricultural Development*, p. 131.
36. Murray, *Modern Iran*, p. 69.
37. English, P.W. *City and Village in Iran* (Madison: Wisconsin University Press, 1966), pp. 135-6.
38. *Salnameh-e Amari 1361* (Tehran: Iranian Statistical Centre, 1362, 1973/74), p. 358.
39. Smith, A. *Blind White Fish in Persia* (London: Readers Union/Allen & Unwin, 1954), p. 181.
40. English, *City and Village*, p. 139.
41. Ibid., pp. 139-40.
42. Smith, *Blind White Fish*, p. 79.
43. English, *City and Village*, p. 50.
44. Bonine, M.E., 'The morphogenesis of Iranian cities', *Association of American Geographers, Annals*, 69/2 (1979), pp. 208-24.
45. Planhol, X. de, 'Geography of settlement' in Fisher, *Cambridge History*, p. 431.
46. Ibid., p. 436.
47. Ibid., p. 431.
48. De Planhol described a zone lying from eastern Azarbayjan through Kermanshah to Fars as the core area of survival of strip farming. Ibid., p. 431.
49. Fischer, M. 'Persian Society: Transformation and strain' in Amirsadeghi, *Twentieth Century Iran*, p. 183.
50. English, *City and Village*, p. 139, quoted the one-kilometre Hujjatabad *qanat* in the Kerman region as taking 27 years to complete, which, although rather more time-consuming than average, gave an excellent reason why

groups and individuals could not walk away lightly from their dependence on, and responsibilities for, the communal water supply systems.

51. Hazrat Aqa, a religious notable of Behdokht in Khorasan, spent very large sums of money, reputedly more than a quarter of his wealth, over a 30-year period between the 1930s and the 1960s. in the improvement of *qanats* in and around his lands surveyed by the author during fieldwork at that time.

52. *Juis* were channels fed along the contour from a river or stream offtake to provide in most cases a single water supply for a village. Among the most developed of *jui* systems was that on the Hari Rud in in Afghanistan. See Ercon Ltd *Hydrogeological Investigation of the Hari Rud Basin*, Vol. 1, General Report (Bracknell, 1974), pp. 39-41.

53. In Tehran, for example, the main water supply until recent years came from some 34 *qanats*.

54. Overseas Consultants Inc., *Report*, Vol. III, Exhibit D-12.

55. *Inter alia*, see Beaumont. P., 'Qanat systems in Iran', *International Association of Scientific Hydrology, Bulletin*, 16/I (1971), pp. 39-50; English, P.W. 'The origin and spread of qanats in the Old World', *Proceedings of the American Philosophical Society*, 112/3 (1968), pp. 170-181; Goblot, H. *Les Qanats: Une Technique d'Acquisition de l'Eau*, (Paris: Mouton Editeur, 1979); Hartl, M. *Das Najafabadtal: Geographische Untersuchung einer Kanatlandschaft im Zagrosgebirge (Iran)*, (Regensburg: Regensburger Geographische Schriften, Selbstverlag, 1979) and Kielstra, N.O. 'Ecology and Community in Iran', unpublished Ph.D. thesis, University of Amsterdam, 1975.

56. The close inter-relationship of *qanats* and the communities they support is not new. Honary's thesis of 1980 suggested that distinctive *qanat* ecosystems could be discerned in the Khur area of the central Dasht-e Kavir. While implicitly accepting the theme that the *qanat* system comprises a human culture as well as a physical ecosystem, Honary did not explicitly articulate this latter aspect in his exceptionally detailed and thoroughly researched study. Honary, 'Qanat and Human Ecosystems in Iran'.

57. Women who conceived out of wedlock or were found guilty of adultery were forced out of their villages and provided a constant flow of unfortunates for the prostitution trade in Tehran and other cities. At the other extreme, the author was well acquainted with a prosperous garage owner-manager in Gonabad who had been converted to Christianity in the American hospital at Mashhad. His life was made extremely difficult. He was forced to divorce his first wife and was proposing to leave the town to escape overt disapproval for his abandonment of Islam.

58. Cf. the example of Abuzaydabad cited by Mahdavi, M. 'A Geographical Analysis of the Rural Economy in the Margins of the Dasht-e Kavir, Central Iran: A Case Study of the AbuZaidabad Area', unpublished Ph.D. thesis, Durham University, 1983.

59. Ibid., p. 140, i.e. *qanats* that received no care and maintenance would theoretically dry up after approximately 30 years.

60. See for example, Plan Organization, *Economic Social and Cultural Land Development Plan*, p. 82.

61. Overseas Consultants Inc., *Report*, p. 150. See also Goblot, *Les Qanats*, pp. 37-44.
62. At Bushruyeh near Ferdaws in Khorasan the over-pumping of the local aquifers was remarked on by pump owners in the mid-1960s during a field visit by the author.
63. English. *City and Village*, p. 139. The hoops were called *nays* and varieties of full and half hoops were employed to keep the flow channel open through preventing roof falls and internal slumping. The fact that the use of *nays* could inhibit water seepage through the stream bed was generally very much secondary to its use a roof lining.
64. Ibid, p. 138.
65. A typical view of the period was expressed by Dr Banki, Minister of State for the Plan and Budget Organization in November 1982. 'The expansion of agriculture in a pivotal manner is quite evident. Imam Khomeini, Ayatollah Montazeri, President Khamenei and the Ministerial Council all hold the view that in case we want to taste independence, we should produce our own food requirements such as bread, meat, etc... our agriculture has been neglected in the past', *IPD*, Tehran, 30 November 1982, p. 11. With specific respect to *qanats*, it is worthy of note that the revolutionary government allocated 4,000 million rials for repair of 2,000 of them shortly after assuming power in July 1979, *Echo of Iran*, Tehran, 12 July 1979, p. 7.
66. See English, *City and Village*, pp. 139-40, for an estimate of construction costs of a three-mile *qanat* at approximately $11,000 per kilometre in US dollars of 1963.
67. Alamouti, A.M. 'Ground water resources of Iran' in *Hydrology and Water Resources Development* (Ankara: CENTO, 1966), p. 427.
68. Overseas Consultants Inc., *Report*, p. 150.
69. English, *City and Village*, p. 139.
70. Plan Organization, *Fourth Plan*, p. 145.
71. Overseas Consultants Inc., *Report*, p. 159. These were themselves merely an elaboration of the work of an anonymous British engineer in 1944, ibid., pp. 180-1.
72. See IBRD, *Economic Situation and Prospects*, pp. 26-7, for a contemporary evaluation of the proposed scheme.
73. Details of the dam and its construction were published in Khuzestan Water and Power Authority *A Commemorative Booklet* (Tehran: Bank Melli Press, March 1963).
74. Development and Resources Corporation, *The Unified Development of the Natural Resources of the Khuzestan Region* (New York, 1959).
75. Nattagh, *Agriculture and Regional Development*, pp. 43-50.
76. See also Ehlers, E. and Goodell, G. *Traditionelle und Moderne Formen der Landwirtschaft in Iran* (Mahburg/Lahn: 1975), pp. 215-23, for an excellent critique of the effects of the modernization of irrigation on traditional agrarian society. Goodell also wrote at length on her research in two villages in the region in *The Elementary Structures of Political Life — Rural Development in Pahlavi Iran* (Oxford: Oxford University Press, 1986). The latter looked in detail at the social and related political effects of the modernization process in one rural settlement outside and one inside the development area

in Khuzestan.
77. Nattagh, *Agriculture and Regional Development*, pp. 46 and 49.
78. Murray, *Modern Iran*, p. 71.
79. Plan Organization, *Fourth Plan*, p. 145.
80. Ibid., p. 156.
81. Aresvik, *Agricultural Development*, p. 66.
82. Mahdavi, 'A Geographical Analysis', p. 143, noted the high incidence of failed wells in Abuzaydabad.
83. In some instances pumped water was fed as an auxiliary supply into the *qanat* stream and distributed through its channels, in which case the new water supply actually strengthened the traditional system.
84. Salmanzadeh, *Agricultural Change and Rural Society*, pp. 243-51.
85. Fisher, W.B., 'The personality of Iran' in Fisher, *Cambridge History*, pp. 735-6, stated that 'The pattern of existence sometimes termed "the great hydraulic civilization" has no place in Iran. Yet the scope and need for water management remains..and are achieved through the *qanat*. Such usage of natural water supplies produces a garden-like pattern of cultivation with very few of the larger units that characterize many other irrigated parts of the world.'
86. Addison, H. *Land, Water and Food* (London: Chapman & Hall, 1955), pp. 83-6, with reference to Egypt.
87. Planhol, 'Geography of Settlement', p. 435.
88. In Gonabad the author was told on several occasions by farmers that some *qanats* had been destroyed by the Mongols, who dropped boulders down shafts to block the *qanat* flow but failed to damage all of them. The persistence of this story is an interesting reflection of the perceptions of the durability of the *qanat* system even against the attacks of the worst of historic enemies.
89. Gischler, C. *Water Resources of the Arab Middle East and North Africa* (London: Menas Press, 1979), pp. 82-4. Gischler was concerned at the rapid draw down of the water table in Qatar and other arid areas of the region through private pumping for agriculture.
90. Such a situation was already the case by the mid-1980s as a result of the high birth rate in towns and the extremely youthful age structure of the urban population.

Chapter 6 Change and Development in the Countryside: Land Reform and Centralization

1. Lambton, *Persian Land Reform*, p. 55
2. Kamaly, 'Iran's experience in agricultural planning', p. 25.
3. It was reported to the author during a visit to the Dusaj area, Saveh, in 1963 that European aid teams had stumbled across small villages (*kalatehs*) hit by the major earthquake in that year the locations of which had been unknown previously.
4. The provisions of the 1960 land reform law were a case in point, where ceilings on individual land ownership were established that applied equally to heavy rainfall and well irrigated areas such as the Caspian coastlands as to the

arid and poor sub-surface water supply area of the south east.
5. Lambton, *Persian Land Reform*, p. 61.
6. Professor Lambton quoted both Arsanjani's 1946 speeches on land reform and his articles in the *Darya* newspaper of 15 and 25 January 1951, that called for the transfer of land to those who cultivated it. ibid., pp. 62 and 50, respectively.
7. Rassekh, S. 'Social Conditions in Iran' unpublished mimeographed report, Economic Bureau of the Plan Organization, p. 4, quoted by Elm, 'Need for industrial changes', p. 3.
8. Hadary, 'The agrarian reform problem', p. 11.
9. Khalil Maleki is put forwards as an example of a socialist thinker who pressed for reform through nationalization of land and water, see Katouzian, *Political Economy*, p. 237.
10. Cf. Pahlavi, *Mission for My Country*, pp. 195-216.
11. Lambton, *Persian Land Reform*, pp. 58-9.
12. Katouzian, *Political Economy* p. 226. 'The land reform did not lead to the foundation of a bourgeois democracy; it resulted in the consolidation of total bureaucratic power over all social classes.'
13. Halliday, *Iran: Dictatorship and Development*, p. 103. 'Although it (the land reform) was carried out under the slogan of "land to the tiller", and was thereby supposed to be egalitarian in character, the implementation of the reform has been such as to create new social divisions in the countryside, in fact to create a capitalist class structure in place of the earlier pre-capitalist one.'
14. Mahdavy, H. 'The coming crisis in Iran', *Foreign Affairs* (October, 1965) pp. 144-6.
15. Pahlavi, *Mission for My Country*, p. 30.
16. Iran was beset by severe balance-of-payments problems in the period 1959-61, which led to the imposition of an economic stabilization programme of considerable severity.
17. Mahdavy, 'The coming crisis', pp. 145-6.
18. Lambton, *Persian Land Reform*, p. 56.
19. Outlines of the 1960 land reform bill are given in ibid., pp. 56-8.
20. At that time the Land Reform Organization estimated that there were 48,592 villages. This assessment was later changed and the conventional figure of 60,000 villages came into common use.
21. For further details of this legislation see McLachlan, 'Land Reform', pp. 692-6.
22. The materials in Persian ranged from straightforward publication of the law such as *Qanun-Eslahat-e Arzi* (Esfahan: Ketabforushi Shahsavari, 1339) to more glossy propaganda books such as *Eslahat-e Arzi dar Iran* (Tehran: Ministry of Agriculture, Esfand 1340). There was also a considerable volume of writing against the reform by the landlords and others.
23. Cottam, *Nationalism in Iran*, pp. 316-17.
24. Katouzian, *Political Economy*, p. 301.
25. One example will indicate the lengths to which the regime went to change the record even in English language texts. The author wrote a short article on land reform in the 1960s for Volume 1 of the *Cambridge History of*

26. See 'Iran's social revolution', in *Iran* (London: Focus Research, 1973), pp. 49-65.
27. Denman, *The King's Vista*, p. 210.
28. Ibid., p. 116.
29. Mahdavy, 'The coming crisis', p. 142.
30. Lambton, *Persian Land Reform*, pp. 359-66.
31. Stobbs, C.A. 'Agrarian Change in Western Iran: A case study of Olya sub-district', unpublished Ph.D. thesis, London University, 1976. Stobbs' thesis showed a growing adverse impact of uncertainty over agriculture and economic counter-attractions to village life from the urban areas in the Olya sub-district of Nahavand, western Iran.
32. Lambton, *Persian Land Reform*, pp. 146-7.
33. A statistical review of the position was given in McLachlan, 'Land reform', p. 687 and an alternative assessment which stresses the dominance of the large landlords was published in Denman, *The King's Vista*, pp. 39-43. Halliday, *Dictatorship and Development*, pp. 117-18, emphasized the position of power of the major landowners such as the Alams and their survival despite the effects of the land reform.
34. See 1339 Agricultural Census (in Persian) Vol. 1, Kol-e kashvar.
35. Lambton, *Landlord and Peasant*.
36. Safi-Nejad, J. *Boneh qabl az eslahat-e arzi* (Tehran: Tus Publications, 1974) and *Asnad-e bonehha* (Tehran: University of Tehran, 1978).
37. Ehlers, E. 'The Iranian village: A socio-economic microsm' in Beaumont and McLachlan, *Agricultural Development in the Middle East*, pp. 151-70.
38. Amini, S. 'The origin, function and disappearance of the collective production units in rural areas of Iran', *Iranian Economic Review*, 5-6 (1978), pp. 145-63.
39. Ehlers, 'The Iranian village', pp. 160-4.
40. Lambton, *Persian Land Reform*, p. 25.
41. Halliday, *Dictatorship and Development*, pp. 116-17.
42. On the Qazvin Plain project, for example, there were 600 *qanats* in use immediately preceding the 1962 earthquake. These were to be replaced by 250 diesel pumped wells under the development programme.
43. Ghahraman, *Water Rights*, p. 24.
44. Mehrad, M. 'Agricultural credit in Iran' in *Cento Seminar on Agricultural Planning*, (Tehran: CENTO, 1972), p. 224.
45. Radji, M.J. 'An Analysis of Wholesale Fruit and Vegetable Marketing in Tehran', unpublished D.Phil. thesis, Oxford University, 1978.
46. Central Bank, *Annual Report*, 1978, p. 101.
47. Agri-business was supported by the Development and Resources Corporation founded in 1955 by two former chairmen of the Tenessee Valley Authority and Lazard Frères & Co. of New York.
48. Development and Resources Corporation, *Agribusiness Opportunities in the Khuzestan Province of Iran*, undated brochure.
49. Nattagh, *Agriculture and Regional Development*, pp. 42-58.

50. Salmanzadeh, *Agricultural Change and Rural Society*, pp. 241-50.
51. Defined as pre-rent surplus, see Bookers Agricultural and Technical Services Ltd. and Hunting Technical Services Ltd., *National Cropping Plan*.
52. Naraghi, 'Rural co-operatives', p. 43, estimated that only 30 per cent of all credits came from the co-operatives as against 33 per cent raised by occasional off-farm work, 32 per cent from money lenders, *pilevars* and merchants and 5 per cent from family and friends.
53. Amini, 'Origin, function and disappearance of collective production units', p. 161.
54. Tapper, *Pasture and Politics*, pp. 23-4.
55. Lambton, *Persian Land Reform*, p. 113.
56. IBRD, *World Development Report*, 1981, p. 179.
57. Amini, 'Origin, function and disappearance of collective production units'.
58. Ehlers, 'The Iranian village', pp. 163-4.
59. Lambton, *Persian Land Reform*, pp. 347-56.
60. See also the debate in Amini, 'Origin, function and disappearance of collective production units.', pp. 146-8 and Safi-Nejad, *Asnad-e bonehha*.

Chapter 7 Agriculture, Oil and Development Planning

1. Keddie, N.R. 'Introduction' in Keddie, N.R. (ed.) *Religion and Politics in Iran* (New Haven: Yale University Press, 1983) p. 11 and Tabari, A. 'The role of the clergy in modern Iranian politics' in ibid., p. 47 and pp. 59-61.'
2. Bonine, E.M., 'Shops and shopkeepers: Dynamics of an Iranian provincial bazaar' in Bonine, E.M. and Keddie, N.R. (eds) *Modern Iran* (Albany: University of New York, 1981), pp. 234-7.
3. Iranian ideas on planning were not lacking at that time but the first commissions for planning studies went to the US concern Morrison-Knudsen. Later Overseas Consultants Inc. took over and prepared their comprehensive study, *Report on Seven Year Development Plan*. Their services were dispensed with following the failure of the US Government to make substantial loans to Iran to fund the proposed plan.
4. Gittinger, *Planning for Agricultural Development*, pp. 87-9.
5. Among the early sections of this study was Bookers and Hunting, *National Cropping Plan*.
6. See, for example, *Iran's Fifth Plan* (Tehran: Kayhan, 1973), p. 10, 'The broad aims of the plan are to...create sources of wealth for the post-petroleum era..'.
7. Baldwin, *Planning and Development*, pp. 113-14.
8. Gittinger, *Planning for Agricultural Development*, p. 90.
9. A flavour of the complexity of the interplay of oil and politics is given in Ferrier, R.W. *The History of the British Petroleum Company*, Vol. 1 (Cambridge: Cambridge Univerity Press, 1982), chapter 13, pp. 588-631.
10. Clawson, P. 'Capital accumulation in Iran' in Nore, P. and Turner, T. (eds) *Oil and Class Struggle* (London: Zed, 1980), pp. 151-2.
11. Jazani, B. *Capitalism and Revolution in Iran* (London: Zed, 1980), pp. 90-4.

12. Mahdavy, *Patterns and Problems*, pp. 428-67.
13. Katouzian, *Political Economy*, pp. 242-50.
14. For discussion of a classification of oil economies based on the degree to which they can sustain exports into the 1990s see Fesharaki, F. and Moussavar-Rahmani, B. *Opec and the World Oil Outlook* (London: Economist Intelligence Unit, 1983), pp. 28-9.
15. Katouzian, H. 'Political economy of development in the oil-exporting countries: An analytical framework', unpublished paper, University of Sussex, 1976.
16. McLachlan, K.S., 'Natural resources and development in the Gulf states' in Niblock, T. (ed.) *Social and Economic Development in the Arab Gulf* (London: Croom Helm, 1980), pp. 88-93.
17. Stauffer, T.R. and Lennox, F.H. *Accounting for "Wasting Assets": Income Measurement for Oil and Mineral-exporting Rentier States* (Vienna: OPEC, 1984).
18. El Mallakh, R. *Kuwait* (Chicago: Chicago University Press, 1968), Chapters 3 and 4.
19. British Petroleum 'Economic development in Iran' unpublished seminar paper, London 1977, p. 3.
20. Only $25 million was granted to Iran by the USA, a mere fraction of the sum sought.
21. Oil output in Iran rose by an annual average of 13.8 per cent, compared with Saudi Arabia 11.0 per cent, Venezuela 2.8 per cent, Indonesia 7.3 per cent and Iraq 5.0 per cent. Of the OPEC members producing throughout the 1960s, only the African states, rising from low base levels, achieved higher rates of growth than Iran.
22. Kamaly, 'Iran's experience in agricultural planning', p. 4.
23. Capital formation advanced rapidly and Gross Domestic Fixed Capital Formation in agriculture rose from $105 million to $305 million in real terms over the fourth plan, at an average annual rate of 23.7 per cent, according to the Central Bank. One third of all investment was in machinery.
24. Clark, B.D. 'Changing population patterns' in Clarke, J.I. and Fisher, W.B. (eds) *Populations of the Middle East and North Africa* (London, University of London Press, 1972), pp. 68-96.
25. Some increase in inflation became apparent in 1972/73 as a result of imbalances in the economy and a proportion of imported inflation as external dependence grew.
26. Hooglund, *Land and Revolution*, p. 86.
27. In the early years, i.e. before 1973, some farm corporations showed signs of commercial success, albeit with considerable hidden government subsidies. A favourable view of their performance at that time was given by Khatibi, N., 'The development of Garmsar farm corporation: A case study', *Oxford Agrarian Studies*, IV/1 (-1976), pp. 1-17.
28. This assumes that the increase in population numbers in the countryside through natural growth, at a rate of some 2.9 per cent annually, was lost to the towns throughout the period.
29. Clark, 'Changing population patterns', pp. 83-90.
30. IBRD, *World Development Report, 1983* (Oxford: Oxford University Press, 1983), p. 191.

31. Pesaran, M.H., 'Income distribution in Iran' in Jacqz, J.W. (ed.) *Iran; Past, Present and Future* (New York: Aspen, 1976), p. 274.

Chapter 8 Agriculture, Revolution and the Rural Community

1. Keddie, N. 'Religion, society and revolution in modern Iran' in Bonine and Keddie *Modern Iran*, pp. 37-58; Keddie *Religion and Politics in Iran* and Bakhash, S. *Reign of the Ayatollahs* (London: Tauris 1985), especially Chapter 1, 'The collapse of the old order'.
2. Abrahamian, *Iran Between Two Revolutions*.
3. Katouzian, *Political Economy*, 1981.
4. Pesaran, M.H. *The System of Dependent Capitalism in Pre- and Post-Revolutionary Iran* (Cambridge, Department of Applied Economics, Economic Reprint No. 64, 1983), pp. 509-11.
5. Hooglund, *Land and Revolution in Iran*, pp. 138-52.
6. Lambton, *Persian Land Reform*, p. 112.
7. The 'great civilization' was a phrase used by the shah in the heady days following the oil boom of the 1970s to describe Iran as it would become once modernization had been completed.
8. Ayatollah Khomeini, 'The granting of capitalory rights to the US' in Khomeini, *Islam and Revolution*, p. 185.
9. Attacks on the land reform came mainly from those who had little understanding of agriculture and its problems or who had personal and political interests to pursue.
10. Khomeini, *Islam and Revolution*, p. 185. 'God knows that I am aware of (and my awareness causes me pain) the remotest village and provincial town...'
11. Afshar, H. 'An assessment of agricultural development policies in Iran' in Afshar, H. (ed.) *Iran: A Revolution in Turmoil* (London: Macmillan, 1985), p. 76, 'The 1978 revolution was essentially urban and the mass of government support is in the large cities.'
12. Hooglund, *Land and Revolution in Iran*, pp. 142-8.
13. Khomeini, *Islam and Revolution*, p. 285.
14. Quoted in *Iran Economic Survey*, No. 221, 12 March 1979, p. 8.
15. In the first budget covering the period from the date of the revolution to the end of the Iranian year in March 1979, agriculture received a mere 8.4 per cent of development funds. Despite an improvement in its position in the year 1979/80, the following year saw a further dramatic 30 per cent fall in the budget.
16. The Mojaheddin-e Khalq is the Organization of the Iranian Peoples' Freedom Fighters, a guerrilla group formed in 1965 with strong affiliations to the Islamic-oriented section of the National Front. It became an important opponent of the shah and later the Islamic Republican regime.
17. *Guardian*, 30 July 1979.
18. The text of the plan indicated that 'acquiring independence is one of the most essential goals and objectives of development (so that) agriculture will enjoy a special significance in development'. *IPD*, 18 January 1983, p. 11.
19. Ibid., 8 February 1983, p. 8.

20. The oil sector was forecast to grow from 20 to 26 per cent of GNP over the plan period. But, to its credit, the plan did propose a reduction in the share of services from 46 to 38 per cent.
21. The Minister of Agriculture lodged a protest at the poor level of finances allocated to his ministry in 1979. See *Echo of Iran, Economic Survey*, No. 700, 14 July 1979.
22. During the author's visit to Iran in 1983 there were several references by villagers in the Tehran region to rivalries between the organizations at the local level.
23. The records of actual development disbursements by the Ministry should not be accepted without the qualifications that (i) funds for development were transferred into the current budget and (ii) the rate of inflation was such as to reduce significantly the value of expenditures over the period.
24. There was a need for direct government allocation of funds to the co-operatives which could not be fulfilled under the existing law. See the debate by Fuad Karimi, deputy chairman of the plan and budget committee of the *majles* in *IPD*, 2 July 1985, pp. 10-11.
25. Total credits allotted through the ministry in 1983/84 were reported at 16,975 million rials.
26. The apparent grant of title to land under phase three of the shah's reforms was a cynical device designed to open the way to farming through centralized farm corporations rather than a genuine move to widen the scope of peasant proprietorship.
27. Gharatchehdaghi, C. *Distribution of Lands in Varamin* (Opladen: C.W. Leske, 1967).
28. A report by the Bonyad-e Mosta'zafin showed that there were eighteen holdings of agricultural land, amounting to 5,448 hectares in the Gorgan and Gombad area belonging to senior members of the former regime which had been sequestrated.
29. See *Iran Economic Survey*, No. 258, 10 December 1979, pp. 4-5.
30. The Fedayin-e Khalq, formed in 1971, was an active guerrilla group which believed in armed struggle against the regime of the shah. It was political rather than Islamic in nature and had a firmly marxist orientation. After the fall of the shah the organization eventually went into hiding and attempted to continue its armed struggle.
31. Reza Esfahani was reported in the Persian language newspaper *Bamdad* to this effect on 31 December 1979 under the heading 'Revolutionaries' programme for land reform', reported in *Iran Economic Survey*, No. 262, 6 January 1980.
32. The Izadi proposals of September 1979 effectively accepted some of the peasant land seizures during the revolution but left the commercial farming structure undamaged.
33. Regional Councils were set up by the landowners, which, supported by the bazaar merchants and others, put up a strong case against the reform programme in the press, the *majles* and in representations to Ayatollah Khomeini.
34. The Jihad-e Sazandegi was highly regarded for its construction works such as making up rural roads, building schools and improving irrigation

water supplies. It was only where personnel from the Jihad put themselves in positions of authority and acted as agents of the central ministerial authority that they fell into bad odour with the local farming community.

35. Husain Mahdavy, a long-standing observer of land reform in Iran, wrote his estimates in the *Ketab-e Agah Book on Iran and the Middle East* (Tehran, 1983), reprinted in *Iran Press Digest*, 24 April 1983, p. 10.

36. Grand Ayatollah Kho'i of Karbala — possibly the most senior of Shi'ite clerics — issued a statement in December 1985 condemning arbitrary land sequestrations. This strongly reinforced the position of Iranian clerics who stood against *band-e jim*. See *Le Monde*, 12 May 1986, though the Ayatollah's representatives later appeared to retract the statement.

37. A review of the amendments to the reform made in 1982 was contained in an interview with Hojjat al-Eslam Fazel Harandi, the chairman of the agricultural commission of the *majles*, *Ettela'at* 12 September 1982.

38. The statement from which this summary was drawn was made but unattributed at the November 1984 seminar and reported in *IPD* 12 November 1984, p. 3. 'If we (those working in and for the seven-man teams) leave, it would be taken as an escape, which is not the case. We guaranteed and have stood by it (the March 1980 law) since the beginning but we have no grounds for work now. We do not want to be apologists for others when we are in the countryside. Category A faces problems wherever it is raised now. Category B is held by the (Mosta'zafin) Foundation and we have been told not to touch Category C (*band-e jim*).'

39. Cf. The Central Bank of Iran's statement 'The main factors responsible for the reduction in the production of major farm crops during the years 1362 (1983/84) were difficulties such as indecision about (the) land ownership situation...', *Economic Report and Balance Sheet 1362*, p. 20.

40. *Ettela'at Supplement* of 13 and 20 April 1986, quoted in *IPD*, 29 April 1986, p. 13.

41. The *boneh* or joint cultivation group was adopted as the mode of collectivization in some districts, taking advantage of the provision in the March 1980 law under Article 3 which called for land to be distributed to partnerships, co-operatives and people sharing land in common.

42. In Egypt immediately before the land reforms 94 per cent of the farmers held less than 2 hectares, while large landowners comprised 4 per cent of the total but held 34 per cent of land. See Warriner, D. *Land Reform and Development in the Middle East* (London: RIIA/Oxford, 1962), p. 30. In Iraq 27 per cent of landowners with less than 2.5 hectares possessed 3 per cent of the land and large owners with more than 50 hectares, accounting for 1.6 per cent of all owners, controlled 26 per cent of land. See Marr, P. *The History of Modern Iraq* (Boulder/London: Westview/Longman, 1985), p. 278.

43. Ashraf, A., 'Peasants, land and the revolution' in *Ketab-e Agah Book on Agrarian and Peasant Issues* (Tehran, 1982) serialized in *IPD*, 10 October 1984, pp. 10 and 11.

44. This latter calculation by the author assumes that the average farm in the 10-50 hectare size group was 17.9 hectares, of which an average of 7.9 hectares would be lost per farm to give 3.1 million hectares at risk. When allowance is made for transfers of 500,000 hectares of land by large landown-

ers to their children under the allowances provided by the amendments to the original land reform legislation, the figure could come down to some 2.5 million hectares. This should be added to the 2.57 million hectares implicitly included in the assessment given by Ashraf noted above.

45. Ayatollah Khomeini's talk to government officials was broadcast on Iranian radio and widely reported in the Persian language press.

46. The radical members of the regime were far from beaten in this or other matters relating to the economy. Ayatollah Khomeini announced on 7 January 1988 that Islamic government could override religious law and practice, on which basis the land reform laws could be resubmitted to the Council of Guardians and, by definition, endorsed.

47. Hojjat al-Eslam Rafsanjani's speech was reported by Iran Radio in *Survey of World Broadcasts* (ME/8285-6/i, 16 June 1986).

48. See *IPD* of 22 July 1986, p. 2. The *IPD* commentary made the point that '...it is easily understood that proposals and recommendations are not merely and by themselves sufficient and effective means to solve the problems and to achieve the desired goals. Repetition of the tiresome slogan "encouragement and improvement of agriculture is necessary" again and again causes misunderstanding and nothing else.'

49. The Jihad-e Sazandegi, it is true, undertook a large number of smaller public works to assist agriculture, for which lists of achievements were published from time to time. For the villagers actually affected by irrigation or road projects completed by the Jihad-e Sazandegi, its work was most important. But, with a limited budget, a loss of dynamism after the first two years of activity, the need to use the Jihad-e Sazandegi to undertake war-related schemes and a growing shortage of foreign-exchange resources, the organization was unable to alter the basic physical realities such as poor water supply.

50. Gross Domestic Product for agriculture divided by the number of rural residents was used to arrive at this rough conclusion since other statistics were entirely absent. Using constant prices of 1353 (1974/75), incomes in the agricultural sector rose from an average of 17,680 rials in 1978/79 to 20,725 rials per head of the total rural community in 1983/84. In contrast, there was a fall in average non-agriculturally sourced GDP from 156,620 rials in 1978/79 to 139,840 rials in 1983/84.

51. The calculation of average incomes must of necessity be somewhat rough. Total rural income was calculated on the basis of the value of agriculture at constant prices of 1353 in GDP for the respective years 1978/79 and 1983/84, less the proportion of agricultural GDP accruing to the urban resident population (2 per cent in the former year as computed from the official report on incomes for 1978/79 contained in the *Scientific Ettela'at* and reprinted in *IPD* of 12 August 1986, pp. 10-11 and an arbitrarily chosen 5 per cent in the latter year), which gave some weight to the rising numbers of rural migrants underemployed in the towns and the poor level of employment in non-agricultural activities in urban areas. It was assumed that receipts from agriculture provided 32 per cent of total rural incomes in 1978/79 and 33 per cent in 1983/84, once again, the proportions chosen, respectively, from the Incomes Survey of the Statistical Centre for the former year and an author's

estimate for the latter. These proportional figures were grossed up to give a total share of the rural sector in national income (GDP). For the sake of consistency, urban incomes were treated as residuals of total GDP. Per capita incomes were constructed by dividing the shares in GDP by the population numbers declared by the Statistical Centre, see Central Bank of Iran, *Annual Report and Balance Sheet*, various years, 'Active population by urban and rural areas'.

52. Some commentators have mistakenly, it is suggested, taken the figures provided by the Statistical Centre surveys to indicate that the sole income from agriculture was that in the self-employed agricultural category. Cf. *IPD* 12 August 1986, p. 10.

53. Sample survey of the Iranian Statistical Centre for the year 1363, published in *Salnameh-e Amari* (Tehran, 1363), p. 85.

54. Iran Statistical Centre sample survey, *IPD*, 2 November 1982, p. 10.

55. See *Jomhuri-ye Eslami* newspaper, 13-14 October 1985. The commentary noted that 'As for underemployment, after the revolution this feature occurred in manufacturing plant as well as the villages, where it happens as a result of environmental causes and agricultural conditions.'

56. The growth rates were 1.9 per cent in rural against 4.4 per cent in the urban areas for 1983/84 according to the Iran Statistical Centre. See Central Bank of Iran, *Economic Report and Balance Sheet 1362*, p. 148.

57. 'The rate of annual growth of the urban population as the result of migration for 1360-65 (1981/82-1986/87)... has been calculated at 1.9... per cent and the total migration at 1.59 million persons,' *IPD*, 10 June 1986, p. 7.

58. Husain Mahdavy took this approach in 'Considerations on the agrarian question in Iran', p. 5. He wrote that 'the rapid rise in the wages in villages had provided the possibility of earning higher incomes through labour... the landless peasants... are better off than the land owning peasants.'

59. Official figures suggested that the proportion of income from agriculture varied between 29 per cent in 1974/75, 38 per cent in 1976/77 and 32 per cent in 1978/79.

60. This ratio was one which became notorious in the interwar years, notably 1921 and 1931, in the principal agricultural exporting countries of the world and was thereafter taken as a signal of conditions of severe economic distress in the countryside.

61. Cf. English, *City and Village*, 1966.

62. Abrahamian, *Iran Between the Revolutions*, pp. 438-9. *Inter alia* Abrahamian noted that 'The bureaucracy so penetrated the rural population that in 1974 the government drew up plans to reorganize the whole countryside', p. 439.

63. See Bellerby, J.R. *Agriculture and Industry: Relative Income* (London: Macmillan, 1956), pp. 37-50, for a thorough statement on this issue. 'In the case of agriculture there may be an intermingling of personal preferences and of social or traditional pressures, inducing men to offer their services more cheaply than in other spheres.'

64. Cf. the debates on this matter by the anthropologist Robert Refield in *The Little Community* and *Peasant Society and Culture* (Chicago: University of Chicago Press, 1963), especially in the latter, pp. 75-7.

65. The *basij* were volunteer military brigades composed of untrained levies who fought as irregulars in the war against Iraq. Most of their recruits were young men under the age of eighteen from the villages. There were many documented cases of young volunteers barely in their teens who left to achieve martyrdom in the war despite the wishes of their families.
66. For a discussion of the reasons why Islam was disposed to be neutral as between urban and other communities see Lapidus, I.M. (ed.), *Middle Eastern Cities* (Berkeley: University of California Press, 1969).
67. 'Disorderly transport of agricultural produce and its consequences' quoted from *Iran Transport Industry Monthly* in *IPD*, 24 July 1984, p. 3.
68. Hojjat al-Eslam Rafsanjani indicated his disapproval of the scale of bazaar manipulation of the market in his speech of 13 June 1986, '...avarice and a temptation to hoard and create black markets for every simple commodity and daily provision still prevails among certain people who wield the consumer market for their own interest', *Survey of World Broadcasts*, p. 5.

Chapter 9 Can Agricultural Self-Sufficiency Be Restored?

1. Cf. Halliday, *Iran: Dictatorship and Development*, pp. 126-7. Professor Halliday suggested that '...oil is now financing the import of food, since (land) reform has failed to increase output'.
2. British Petroleum, 'Economic Development in Iran', unpublished seminar paper, London, 1977, p. 3 and Fig. 1. The BP study argued that rising internal demand for oil and products together with a marginally declining volume of crude oil production could lead to the country losing its role as a principal supplier to the world oil market. This view was not lost on senior Iranian officials with whom the author had discussions at the time.
3. IBRD, *A Study of the Iranian Agricultural Economy* (Tehran: Agricultural Development Bank, 1974), p. 50.
4. Mahdavy, H., 'Consideration of the agrarian question in Iran' in *IPD*, 18 April 1984, p. 12.
5. Ibid., p. 5.
6. Nahavandi, H. and Rad-Serecht, F. 'Le développement de l'économie Iranienne; situation et perspectives', *Iranian Economic Review*, No. 1, Tehran University (1976), p. 1.
7. Among the various debates on this issue see Stevens, P. 'The Impact of Oil on the Role of the State In Economic Development — A Case Study of the Arab World', paper presented at the British Society for Middle East Studies – Middle East Studies Association Conference, School of Oriental and African Studies, London, July 1986, especially p. 8.
8. Yousefi, M. and Abizadeh, S. 'Food self-sufficiency: the case of Iran', *Iranian Economic Review*, No. 5-6, Tehran University (1978), p. 196.
9. Ibid., pp. 201-2.
10. The major works of Jazani were published in a collected version in English in 1980 as *Capitalism and Revolution in Iran*. See the section on agriculture, pp. 46-69, and dependent capitalism, pp. 70-122.
11. Ibid., p. 97.
12. Mahdavy, 'Patterns and problems of economic development'.

13. According to his own witness, René Dumont, an eminent writer of international stature on farm types and world agriculture, wrote to the Iranian Minister of Plans and Budgets, Majid Majidi, in 1977 'l'indépendance économique d'un pays exige un certain degré d'autonomie alimentaire...', cited in Brun, T.A. and Dumont, R. 'Les risques de dépendance alimentaire du modèle de développement agricole en Iran', *Iranian Economic Review*, No.5-6, Tehran University (1978), p. 83.
14. McLachlan, K.S., 'Food supply and agricultural self-sufficiency in contemporary Iran' in *Bulletin of the School of Oriental and African Studies, Vol. XLIX, Part 1 (1986), pp. 149-50.*
15. *Central Bank of Iran Annual Report and Balance Sheet* (Tehran, 1353), p. 71. 'The rise in meat, poultry and fish prices by 25.8 per cent, in fresh fruit and vegetable prices by 25.9 per cent, in the price pf rice by 51.6 per cent, and in dairy products by 16.6 per cent were the contributors to the price index on the food group.'
16. Business International *Iran* (Geneva: Research Report, Geneva, 1984), p. 11. The complete list also included estimates as follows: barley 400,000 tons, maize 400,000 tons, sugar 500,000 tons, butter 60,000 tons, cheese 80,000 tons, powdered milk 40,000 tons, tea 35,000 tons, tobacco 5,000 tons and cigarettes 17,000 million.
17. IBRD, *World Development Report 1985*, pp. 184-5.
18. Statistics for 1983/84 from Iranian Customs Authority reports.
19. Adel, A.H. *Ab va hava-ye Iran*, Vol. 1 (Tehran, Daneshgah-e Tehran, Khordad Mah, 1339) and Ganji, M.H. 'Climate' in *Cambridge History*, pp. 212-49.
20. For a concise account of the climate of Iran see Fisher, W.B. *The Middle East* (London: Methuen, 1963), pp. 289-93.
21. There was a fivefold division of the country proposed by Ganji on the basis of thermal conditions, 'Climate', pp. 227-9, while a fourfold division using thermal/vegetation factors was suggested by Bobek, H. 'Beitrage zur klima-ökologischen Gleiderung Irans', *Erkunde*, 6 (1952), pp. 65-84, including the *sarhad* or the cold highlands, *sardsir* or cool temperate zone, the intermediate warm temperate zone and the *garmsir* or hot subtropical zone. Bobek also laid down a ninefold climatic division in this same review based on rainfall regimes, description of which is given in the well illustrated volume by Ehlers, E. *Iran: Grundzuge einer Geographischen Landeskunde* (Darmstadt: Wissenschaftliche Buchgesellschaft, 1980), pp. 74-81.
22. See Stobbs, 'Agrarian Change in Western Iran', for the Nahavand region.
23. Curzon, *Persia and the Persian Question*, p. 488.
24. Overseas Consultants Inc., *Report*, Vol. III, pp. 60-1.
25. Bonine, 'From *Qanat* to *Kort*', pp. 152-3.
26. As seen on personal field visits to the Dez project area in 1971 and 1978. Complaints that land had not been properly levelled and that elsewhere soils had been damaged in the course of reclamation were made by the Iranian scientists working at the experimental farm at the site.
27. English, *City and Village*, pp. 92-4.
28. Mahdavy, 'A Geographical Analysis of the Rural Economy', p. 388.

29. Ibid., pp. 390.
30. Alessa, S. *The Manpower Problem in Kuwait* (London: KPI, 1981), p. 38.
31. See Lambton, *Landlord and Peasant*, pp. 295-336.
32. Bonine, *Shops and Shopkeepers*, pp. 237-8.
33. A sign of the attractiveness of agriculture as an investment was the appreciable construction of *qanats* at that time. One octogenarian informant, Dr Zabih Qorban, from Abadeh in the Shiraz region told me that his father had dug a new *qanat* every year, each of twenty cultivation areas fed by the new *qanats* being named after one of his family. Funds for the enterprises were easy to raise and returns on investment were good.
34. Central Bank of Iran, *Annual Report,* various years.
35. United Nations Economic and Social Commission for Asia and the Pacific, *Annual Report* (Bangkok, 1983), p. 88.
36. Salmanzadeh, C. 'Aspects of the sugar production and consumption in Iran', *Zuckerindustrie*, 109/11 (November 1984), p. 1020.
37. Professor Lambton noted the effects of improved living standards as early as the first stage of land reform in 1962/63: 'By 1964 there was a noticeable rise in the general standard of living in the north. ...it was particularly marked in the improved diet of the peasants in the villages where the land reform had been operative.' *Persian Land Reform*, pp. 192-3.
38. The estimate of population for 1363 (1985/86) by the Iranian Statistical Centre was 43,414,000 as shown by *Salnameh-e Amar 1363*, p. 56. The growth rate was some 3.19 per cent.
39. In crude terms the author estimated in 1986 that some three-quarters of a million people each year were leaving the villages, on the assumption that the rate of natural increase in population in rural areas ran at 3.3 per cent annually yet the number of persons resident in rural areas remained stagnant to slightly declining.
40. See *Annual Report and Balance Sheet* for 1360, 1361 and 1362.
41. A detailed breakdown of subsidies for 1982/83 showed 37,000 million rials ($411 million) for wheat and 6,000 million rials ($66.5 million) for sugar. A further 27,000 million rials ($300 million) was expended on subsidy of fertilisers used by farmers. See *Annual Report and Balance Sheet*, Central Bank of Iran, 1361, p. 19.
42. In 1983/84 the government purchased 832,000 tons of wheat and 140,000 tons of rice under the guaranteed price scheme.
43. The US Department of Commerce introduced sanctions against Iranian pistachio imports in 1986. See *Middle East Economic Digest*, 15 March 1986.
44. *IPD*, 22 July 1986, p. 2. '...natural resources remain dormant and unexplored... only 20 per cent of fertile lands... have been used...'.
45. Iranian carpet exports were expanded significantly in 1979/80 and 1980/81 from an average of some $100 million/year to more than $400 million/year. Controls on exports brought the value down to $89 million in 1983/84.
46. Katouzian, *Political Economy*, p. 306. 'growing food deficits... were caused by... increases in demand and by the petrolic approach to agriculture'.
47. In the modern period oil rose as a component of GDP from 21.2 per cent

in 1963/64 to 24.7 per cent in 1967/68 (at prices of 1338), 50.5 per cent in 1973/74 and 35.8 per cent in 1977/78 (at prices of 1353).

48. There was an average rise of 6 per cent/year in consumption of food per head as measured by calories between 1972 and 1975, the years before and after the oil boom. Statistics taken from Vahidi, I. 'Iran in the 1980s: Agricultural Outlook', mimeographed paper (Tehran: Institute for International Political and Economic Studies, 1977), p. 11.

49. Clark, 'Changing population patterns', p. 79. Clark's figure was for 1966.

50. The Central Bank of Iran reported in its *Annual Report and Balance Sheet* for 1353, p. 63, that imports of capital goods as a proportion of total imports dropped from 25 per cent in 1972/73 to 20 per cent in 1974/75.

51. For a fuller account of the problem, see McLachlan, 'Disaster of the oil boom', p. 25.

52. Agricultural workers received approximately half the national wage in 1965/66 but this ratio fell to less than 30 per cent by 1975/76.

53. It was estimated that less than a third of all villages had electricity or piped water in 1978. Less than 3,000 villages out of a nominal total of 60,000 had a direct service from a black-top road and less than one per cent of villagers owned a private car, which compared badly with urban areas on all counts.

54. The literature concerned with the characteristics of the oil economy has grown considerably in recent years. Dr Katouzian both originated a typology of oil-exporting states in his paper 'The political economy of development in the oil-exporting countries: an analytical framework' given at Sussex University in 1976 and applied many of these same ideas in his later book *The Political Economy of Modern Iran 1926-1979*, pp. 244-50. The author's own contribution of a model of change in which various categories of oil-exporting states, including resource-rich states such as Iran, developed a dynamic of evolution towards oil dependence at an accelerated pace after 1973 appeared in the *Middle East Annual Review* in 1979, pp. 25-7, and in 'Natural resources and development' in Niblock, *Social and Economic Development in the Arab Gulf*, pp. 86-93. Other written materials that assist in the understanding of the economic and social nature of oil-exporting countries such as Iran include Birks, J.S. and Sinclair, C.A. *Arab Manpower* (London: Croom Helm, 1980), pp. 9-12, Mabro, R. 'Development — Defects in Opec's fast growth strategy', in *Middle East Annual Review 1977*, pp. 23-9, Stauffer. T.W. with Lennox, F.H. *Accounting for "Wasting Assets"*, Pamphlet Series 25 (Vienna: OPEC, 1984) and Stevens, 'The Impact of Oil on the role of the state in economic development'. All the above papers agreed on one theme — that although oil shares some characteristics with other depleting natural resources, its effects on countries with high levels of exports are so powerful as to touch most aspects of economic and indeed social activity within them. The implication, too, was that oil at best was a mixed blessing for all major oil-exporting states, and for the Third World oil exporters created as many, if not more, economic problems than it solved, particularly through the dwarfing and emasculation of the non-oil productive assets within the economy.

55. The total population reported was 48.3 mn, rather higher than mid-year estimates of 46.2 mn for 1986.

56. IBRD, *World Development Report 1985*, p. 211.
57. See the forecasts of Yousefi and Abizadeh 'Food self-sufficiency'; pp. 201-2. The two professors of the University of Shiraz suggested population growth rates ranging between 2.3 per cent and 2.5 per cent for 1980, 1.9 and 2.3 per cent for 1985, and 1.6 and 2.0 per cent for 1990, whereas actual growth was much higher according to the Iranian Statistical Centre — 2.9 per cent in 1980 (*Salnameh-e Amari*, 1361, p. 70) and well over 3 per cent for 1985 based on ISC figures for 1984/85 of 3.2 per cent.
58. In 1983/84 54 per cent of the population was under 19 years of age and 62 per cent younger than 24 years. See *Salnameh-e Amari*, 1363, p. 56.
59. The Minister for Energy, Taqi Banki, announced at a seminar in Tehran in April 1986 that 'Per capita income fell 85,000 rials ($985 at an exchange rate of 86.4 rials to the US dollar) in 1983/84 from 106,000 ($1,500 at a rate of 70.5) in 1978/79.' The minister blamed the deterioration on a 17 per cent rise in population numbers.
60. Joffe, E.G.H. and McLachlan, K.S. *Iran and Iraq: The Next Five Years* (London: Economist Publications, 1987).
61. Taking into account the most likely population forecast of an average rate of growth rising to over 3.5 per cent during the period.
62. Iran's standard of diet was extremely poor in the period 1969-71, with per capita dietary energy supplies only 90 per cent of nutritional requirements according to *Economic and Social Survey of Asia and the Pacific* (Bangkok: United Nations, 1984), p. 88. In fact, more than half of the population was estimated to be undernourished at that time. Political unrest occasioned by food shortages seemed to be a real risk in view of the large numbers of people subsisting on inadequate diets.
63. The author observed in 1964 'Nonetheless, it will be necessary to demonstrate clearly that once the reform is complete, no further radical change in ownership will be introduced, so that in both landlord and peasant areas development can take place in an atmosphere of confidence without the feeling that the land reform could be no more than a further temporary upheaval in the long and disturbed history of landed property in Iran'. 'Land reform', in *Cambridge History*, p. 713.

Index

Abadan 83, 170, 185
Abrahamian, E. 60
Abuzaydabad 78, 95
administration 29, 30, 32, 63, 106, 143, 182
afforestation 99, 100
Afshar 239
agreements
   Anglo-Persian (1919) 30
   bilateral (oil for goods) 175
   Irano-Soviet Joint Co-operation (1966) 164
   Supplemental Oil 165
agribusiness 72, 100, 134-7, 152, 154, 157, 170, 173, 204, 206, 236
Agricultural Co-operative Bank 132-4 *passim*, 147, 148; – Development Bank 106, 133, 134
agro-industry 75, 134-7, 177, 183, 201, 202, 204
Ahwaz 83, 135, 185, 186
aid 49, 52-4 *passim*, 106, 110, 154, 177, 230
Alam, Asadollah 56, 57, 59, 60, 124; family 124
Alborz 11, 12, 15-17, 19, 22, 81, 238
Aliabad *qanat* 80, 91, 93
Amini, Ali 55-6, 113
Amini, S. 126, 144
amenities 136, 146, 218-19, 258, 294n.53
Amuzegar, Dr 177, 178
Andimeshk 99
Anglo-Iranian Oil Co. 38, 41, 47, 165
anti-profiteering 177, 178
Arak 168, 170, 186
Aras dam 135, 172
armed forces 31, 33, 38-40 *passim*, 53, 146, 162
arms supplies 39, 52, 107, 175

Arsanjani, Hasan 55, 56, 107, 113-14, 123, 147-9 *passim*
Arya Mehr farm corporation 177
Ashraf, A. 212
Astan-e Qods 90
Australia 234
Azarbayjan 14, 19, 42, 44, 86, 135, 138, 145, 183, 184
Ayandegan 191

balance of payments 167, 227
Bakhtiari 30, 82, 145, 239
Baluchestan 14, 79, 80, 124, 138, 219
Bandar Abbas 16
Bandar Shahpur (Khomeini) 170, 185, 186
band-e jim 206, 208-13
bandkari 79, 80, 82
barley 87, 88, 247, 249
basij 220, 291n.65
Bayazeh 70, 74
bazaar 35, 44, 53, 59, 132, 134, 153-4, 223, 241-2, 253-4, 291n.68
Bazargan, Engineer 192
Behdokht 87, 98
Beheshti, Ayatollah 207
Bharier, J. 50
boneh 126, 144, 151, 208, 210, 221, 288n.41
Bonine, E.M. 92
Bonyad-e Mosta'zafin 23, 199, 202-3, 209, 212
Borujerdi, Grand Ayatollah 60
Bradshaw, D. 82
Britain 30, 38, 40-2, 47, 52, 110-11, 165
bureaucracy 33, 35, 106, 120, 146, 148, 154-5, 226 *see also* administration

capital formation 179, 182-3, 285n.23
car assembly 168

carpets 81, 238-9, 253, 258, 293n.45
Caspian plain 11, 17-19 *passim*, 30, 33, 37, 84-5, 152, 183
Census, agricultural 66, 89, 122, 210
centralization 31, 62-4, 108, 120, 163, 219, 266
Chupanan 78
Civil Code 35, 69-71, *passim*, 205
climate 9, 11, 14-22, 76-86 *passim*, 151, 234-5, 292n.21
collectivization 202, 208, 221-2, 288n.41
communications 37, 39, 62, 167, 276n.41
Constitution (1906) 28, 29
constitutional movement 7, 27-31, 108
consumption, food 243-6 *passim*, 252, 254, 262-3, 294n.48, 295n.62
contract working 74, 87, 88, 126
co-operatives 62, 115, 120, 128, 131, 133, 134, 136, 140, 146-7, 149, 173, 199, 201, 210, 212, 246
Cottam, R.W. 113
cotton 17, 33, 85, 87, 193, 242, 249, 252-3
Council of Guardians of the Constitution 126, 193, 209, 222
credits 5, 8, 62, 115, 119, 120, 131-4, 137, 147, 155, 192, 197, 201, 223, 241, 284n.52
crown lands (*khaleseh*) 27, 32, 34-5, 43, 48-54 *passim*, 105, 106, 109, 143, 273n.16; Decree for Distribution of (1951) 53
Curzon, Lord 5, 22-3, 27, 36, 236

dairy industry 202
dam construction 6, 17, 20-1, 24, 44, 68, 99-100, 104, 105, 155, 166-8 *passim*, 172 *see also individual headings*
Dariush Kabir dam 172
Dasht-e Gorgan 17, 253; – Kavir 11, 13, 70, 71, 76-80, 95; – Lut 11, 13, 77-9; – Moghan 21, 32, 135, 201, 204
debt 140, 147
decentralization 200-1
defence 156, 160, 175
Deh-e Juimand *qanat* 91, 93
Delui *qanat* 80, 90
dependence, external 5, 191, 226-34, 264
Dez project 6, 20-1, 100, 102, 105, 134, 135, 166, 167, 185, 192, 236, 237
Dez Kar Company 135

Dezful 17, 83
Distribution and Sale of Rented Farms Act (1968) 117-18
diversification
 crop 88, 177, 235-7 *passim*, 240
 economic 63-4, 226, 238-40
drought 27, 31, 63, 82
dryland culture (*daym*) 15, 16, 18, 24, 79-81 *passim*, 86-9
Dumont, René 228, 292n.13

Eastern sumps 79-81
Ebtehaj, Dr 156, 165
education 35, 54, 63, 143, 145-6, 192, 194, 258
EEC 234
Ehlers, E. 126, 147
elections (1963) 56
electricity, generation 20, 68-9, 73, 100, 134, 170; supply 136, 218-19, 258
English, P.W. 89, 91, 92, 98, 239
environment 6, 8, 9, 10-22, 65, 75-86 *passim*, 151, 224, 234-5
Esfahan 11, 14, 77, 90, 168, 170, 184-6 *passim*
Esfahani, Reza 200, 206, 210
expenditure, public 48-9, 164-77, 192, 194-8, 200 *see also* plans
exports 17, 155, 225, 241, 252-4, 256, 258, 263; oil 158, 159, 161, 165, 172, 196, 226, 228, 229

farm corporations 118-20, 128, 133, 136, 137, 149, 150, 152, 157, 170, 172, 173, 177, 183, 201, 204, 206, 212, 285n.27
FAO 106
Fars 75, 118, 121, 138, 144
Fedayin-e Khalq 206, 287n.30
Ferdaws 81, 131, 239
fertilizer 5, 8, 89, 150, 257
Fisher, W.B. 103
foreign exchange 9, 38, 49, 96, 97, 161, 200, 226, 227, 229, 232, 252, 257, 264
foreign policy 162
fruit growing 14, 16, 17, 34, 84-6 *passim*
Full Powers Act (1952) 47-9 *passim*, 105, 106, 274n.3

gavbands 108, 125, 127, 128
GEC 54
Germany 40, 41
Ghaffurabad 78
Ghahraman, F. 70, 89

Index

Gilan 89, 140, 183, 184
Goblot, H. 96
Gonabad 15, 80, 87, 91, 93, 102, 124, 129, 204
Gorgan 15, 85, 87, 88, 183, 184, 193, 204, 253, 287n.28
'growth poles' 136, 155, 173, 186, 188, 258
Gulf coast 16, 83-4

Hadary, Gideon 5, 49, 65, 66
Haft Tappeh 17, 21, 100, 201
Halliday, F. 108, 124
Hamadan 14, 240
harim 70, 96
health 54, 115, 143, 192, 246, 258; Corps 143, 146, 148, 246
Hojjatabad *qanat* 98
housing 136, 175, 194
Hovayda, Amir Abbas 63

IBRD 134, 167, 261
Imam Reza 128, 129
immigration, foreign 182
imports 161, 175-7, 200, 225-34, 256-7
  agricultural 2, 170, 176, 225, 228-34, 242, 246, 257
  food 4, 5, 33, 223, 225-8, 245, 249, 252, 256, 257, 263-6 *passim*
income 184, 213-16, 218, 228, 232, 239, 243, 262, 270n.39, 289n.50, 51, 52, 290n.59,60, 295n.59
industrialization 36-7, 39, 40, 54, 64, 168, 184-6, 227, 246, 267n.9, 273n.25
industry 2, 5, 54, 156, 165, 168, 184-6 *passim*, 227
inflation 42, 50, 167, 168, 175, 177, 226
inheritance 69, 116
innovation, crop 33, 37, 43, 44, 150, 242 *see also* diversification
International Monetary Fund 55
investment 184-6; in agriculture 3, 8, 9, 35, 44, 88, 105, 132, 155, 167, 172, 173, 176, 183, 188, 192, 194-5, 197, 200, 218, 221, 222
iqta system 27-8
Iran America Agro-industrial Co. 100, 135; – California Co. 135; – Shell Cott Co. 135; – Statistical Centre 214, 216-17, 230
irrigation 3, 4, 14, 17, 19, 20-2, 25, 43, 50, 65-105, 131, 134-5, 147, 156, 166, 188, 202, 203, 222, 237, 266; Independent – Institute 43-4, 50, 72, 73
Islam 189-90, 194 *see also* law, Islamic
Izadi, Ali Mohammad 192, 199, 206, 221

Jandaq 78, 240
Jangal 80, 86
Javadieh *qanat* 90, 98
Jaz Murian 14-15
Jazani, Bizhan 228, 291n.10
Jeep 54
Jihad-e Sazandegi 199, 202, 207, 208, 217, 220, 287n.34, 289n.49
Jiroft 15, 21, 135
joint companies 112, 114-15, 118
Juimand 80, 129

Karaj river 105, 166, 185
Karkheh river 83, 84
Karun river 16, 20, 83, 100
Kashan 239
Katouzian, H. 158
Kavir basin and margins 76-7
Kavir-e Zangi Ahmad 13
Kennedy, President 107, 109
Kerman 13, 14, 77, 90, 91, 98, 114, 121, 130, 185, 186, 239
Kermanshah (Bakhtaran) 32, 185, 186, 205
Khamseh 82
Kharg Island 170
Khash plateau 14
Kho'i, Grand Ayatollah 288n.36
Khomeini, Ayatollah 60, 62, 189-91 *passim*, 193, 196, 209-10, 212, 213, 221, 223, 289n.46
Khorasan 11, 13, 80, 81, 86, 90, 124
Khorramabad 82
Khorramshahr 83
khoshneshin 119, 140, 144, 151
Khosrawabad 84
Khur 78
Khuzestan 16-17, 21, 38, 75, 83-4, 134, 135, 184, 185, 204, 230; Water and Power Authority 83-4, 99
Kuchik Khan 30
Kuhgiluyeh 82
Kurdestan 6, 14, 44, 183, 205, 219, 220
Kurdish People's Republic 42
Kurosh Kabir dam 172
Kuwait 161, 163, 240

labour 3, 8, 89, 120, 175, 180, 182, 216, 222, 236, 247

Lambton, A.K.S. 5, 7, 47, 50, 71, 107, 108, 145
land 24, 65-8
  abandonment 8, 22, 23, 83, 222, 257
  ceilings 87, 111-13, 115, 137-8, 281n.4
  reclamation 6, 19, 20-1, 24, 65, 87, 88, 100, 172, 194, 222, 242
  reform 6-8 *passim*, 51, 55, 57-9 *passim*, 61, 72, 87, 88, 94, 108-25, 129, 130, 132, 137-42, 144-5, 147, 148, 150, 154, 167, 168, 187-9, 191-3, 200-13, 225, 243, 265; phase one 23, 58, 111-14, 123, 125, 139; phase two 62, 87, 112-16, 128, 139-42, 149, 265; phase three 117-20, 128, 133, 140, 142, 148, 149, 172, 265; 1980 law 206-13, 217-18, 221-2
  registration 32, 43
  seizure 192, 193, 200, 202, 204-6, 221, 287n.28
  tenure 4, 6, 8, 27-30, 32, 33, 35, 39, 50, 115-16, 200, 203-13, 217-18, 221-2 *see also* ownership, land
Land Reform Council 112
Land Reform Organization 56, 61, 80, 112-16 *passim*, 118, 121, 122, 124, 140, 145
landlords 6-7, 28-9, 35, 47, 48, 51-3 *passim*, 58, 62, 83, 105, 107-9 *passim*, 111, 115, 119, 122-5, 130, 132, 139, 144, 146-51 *passim*, 154, 167, 206, 207, 236, 241, 269n.31; Association 58, 275n.28
Lar 203
law 35, 143
  customary ('urf) 71-2, 205, 277n.10
  Islamic 35, 50, 58, 69-73, 111, 189, 205-6, 208-10
  inheritance 69
  Qanat (1930) 72
  water 69-75
leases 8, 115, 126, 129, 208, 210
Lennox, F.H. 158-9
literacy 57, 115, 146; Corps 63, 143, 146, 148
livestock 14, 23, 34, 66, 68, 82, 232, 238, 240, 249, 272n.34
loans, foreign 166-7
Lurestan 6, 32, 183

Mahabad 42
Mahdavi, M. 78
Mahdavy, H. 110, 119, 157, 208, 227, 228, 239

Makran 14
Manjil 85
Mansur, Hasan Ali 63, 114
marketing 8, 49, 191, 200, 223, 243, 246
Mashhad 15, 16, 59, 128, 129, 184, 186; basin 97-81
Masjed-e Sulayman 159
Mazandaran 35, 43, 89, 183-5 *passim*
meat 231-2, 234, 243, 249
mechanization 9, 43, 87-9, 111, 113, 115, 117, 150, 176, 200, 222, 236, 242
Meshkini Ayatollah, 207
migration 240; rural/urban 3, 9-10, 23, 78, 81, 83, 103, 144, 146, 156, 173, 176, 177, 182, 188, 203, 216-22, 224, 239, 240, 247, 249, 267n.12, 290n.57, 293n.39
Millspaugh, A.C. 31-2
Ministry of Agriculture 24, 192-3, 199, 201-3, 230
modernization 1, 2, 9, 35-9, 44, 57, 64, 88-9, 109-11 *passim*, 116, 150, 153-4, 160, 176, 187, 190, 201, 242, 243
Mohammad Reza Shah 48, 51-64, 107, 113, 158, 162-3, 189, 190, 265, 266
  and land reform 58, 61, 108-11
  and *majles* 51-64
  and modernization 56-8, 63, 110-11, 142-3, 153-4, 157, 162-3, 189, 190
Mojahedin-e Khalq 193, 206, 286n.16
Montazeri, Ayatollah 207
Morrison-Knudsen 164
Mosaddeq, Dr 47-50 *passim*, 105, 107
Muhammad Ali Shah 29-30
Murray, John 5, 89, 100
Musavi, Mir Hossein 209

Naser Khan 58,
National Cropping Plan 66, 67, 155
National Front 29, 47, 49-51, 56, 57, 61, 107
National Iranian Oil Co. 52
nationalism 37, 40, 41, 160
nationalization 220
  forest 57, 143
  oil 1, 165
  water 50, 72-4, 132, 143
Nattagh, N. 100
Negar 90
New Zealand 234
nomads 6, 14, 16, 34, 82, 145, 238, 239

occupation, Allied 41-2
oil 1-2, 5, 9, 38, 41, 47, 52, 103, 155-78,

185, 226-9, 254-61, 294n.54
  crisis (1951-3)   1, 45, 47, 48, 110, 159, 161, 178
  exports   158, 161, 165, 169, 172, 196, 226, 228, 229
  nationalization   1, 165
  prices   163, 218, 223, 227
  revenues   38, 48, 155, 156, 159-63, 165, 167-9, 173-8, 187-8, 196, 223, 226, 229, 249, 254-7 *passim*
Okazaki, S.   53, 87
OPEC   159, 170
Organization for Protection of Consumers and Producers   252
Overseas Consultants Inc.   24, 65, 94, 96, 99, 164-5, 236
ownership
  land   4, 5, 7, 29, 32, 33, 39, 48, 49, 51, 52, 88, 101, 111-12, 114-18, 121-5, 127-9, 144, 148, 149, 154, 187, 192, 199-201, 203-13, 220-2, 295n.63
  qanat   90, 93
  water   4, 69-75, 90, 93, 101, 114, 128, 130, 132
ox-ploughing   89, 108, 126, 238

Pahlavis   1, 4 *see also individual headings*
Pahlavi Foundation   50, 51, 105
Palestine issue   159
pest control   89, 150
petrochemicals   168, 170, 185, 226
de Planhol, X.   92, 103
Plan Organization   19, 48, 49, 66, 154-6, 164-8 *passim*, 182
planning   4, 94, 153-78
Plans
  First   48, 49, 154, 164-5, 228
  Second   21, 49, 51, 55, 68, 105, 155, 166, 228
  Third   20, 21, 68, 132, 155, 167-8, 183, 185, 228
  Fourth   20, 68, 101, 136, 169-72, 183, 185, 229
  Fifth   20, 68, 136, 154, 155, 172-6, 186-7, 229
  Sixth   20, 178
  Islamic Republic   22, 193-6, 261
population growth   5, 54, 66, 68, 98, 103, 170, 224, 228, 243, 246-7, 255, 261-4 *passim*, 294n.55, 295n.57;
  urban/rural balance   183-4, 217, 247, 261, 290n.56
Price, O.T.W.   22

pricing policies   3, 8, 33, 88, 176, 223, 232, 250-4, 257, 266, 269n.37 *see also* subsidies
processing   36, 134-7, 246
productivity   7, 25, 108, 116, 117, 147, 173, 247, 249
profit-sharing, industrial   57, 143
public works   156, 257
Public Ownership Law (1975)   176-7
pumps   20, 22, 70, 74, 78-9, 83, 85, 90, 95-8 *passim*, 101, 103, 104

Qa'enat   79, 124
qanats   4, 15, 17, 20, 22, 44, 69-71, 74, 76-81, 85, 86, 89-98, 101-4, 130-1, 203, 220, 237, 266, 279n.56, 57, 293n.33 *see also individual headings*
  construction   72, 93, 97, 98, 278n.50
  costs   97-8
  hava bin   77
  law (1930)   72
  management   4, 69-70, 90-1, 94, 277n.10
  maintenance   69-72, 80, 90, 96, 97, 131, 147
  ownership   90, 93
  qarq-ab   77
Qasabeh Shahr *qanat*   90, 91, 93
Qashqu'ai   58-9, 145, 239
Qazvin   11, 131, 186
Qom   59

radio   114
Rafsanjani, Hojjat al-Eslam   213, 291n.68
railways   36, 37, 39
rainfall   11, 14-22 *passim*, 75, 77-81, 84-6 *passim*, 130
Rasht   186
rationing, food   249, 252, 262, 264
recurrent costs   182
regional development   183-8 *passim*
Regional Councils law (1937)   72
religious classes   29, 35-6, 50, 51, 53, 58-60, 62, 153-4, 189 *see also 'ulama*
rentier economy   157-8, 228
rents   139, 204
revolution   4, 97, 189-224
Reza Shah   31-44, 71, 72, 153, 165, 242
rice   17, 85, 232, 243, 250, 254, 262
rights
  cultivation (*haqq-e risheh*)   7, 115, 116, 114
  water   69-75, 130, 144

riots (1963)   58-60
roads   36, 37, 39, 54, 258
Ruhani, Ayatollah Sayyed Sadeq   207
Rural Development and Extension
    Corps   143, 146, 148
Russia   30, 38

saffron   81
Safi-Nejad, J.   126
Sai, K.   49, 65, 66
Salamati, Mohammad   200-1
schools   35, 63, 136
Sefid Rud   20, 75, 85, 105, 166, 185
self-sufficiency   1, 33-4, 50, 155, 176,
    178, 191, 223, 230-54, 260-6
Shahdad   13
Shah Abbas Kabir dam   172
Shahpur I dam   172
Shahvur Co.   135
share-croppers   47, 112, 115, 125, 126,
    139, 241, 273n.21
Shiraz   16, 34, 184-6 *passim*
Shuster, Morgan   31
Sistan   15, 32, 43, 124, 138
Sixth of Bahman reforms   61-3 *passim*,
    143, 145-6, 157
Smith, A.   92
social change   94-5, 142-9
soil   9, 12-17, 74-5, 156, 234, 237
Soviet Union   40-3 *passim*, 110, 135,
    164, 168
Stauffer, T.R.   158-9
storage facilities   8, 49, 223 *see also*
    water
structure
    agrarian   4, 6-8, 27-9, 39, 47-8, 91-5,
        107, 108, 115, 118, 121-32, 144-5,
        147-52, 188, 204-13, 221-2, 236, 241,
        243, 266
    village   91-3, 101-2, 104, 237
subsidies, food   200, 214, 232, 249, 252,
    293n.41
Suez crisis   52
sugar   21, 33, 44, 135, 201, 230-2, 234,
    236, 243, 246, 249, 251-4 *passim*, 262
Supplementary Fundamental Law
    (1907)   29
Susangerd   83
stablization, economic   54-5, 177
Syria   8

Tabriz   170, 184-6 *passim*
Tapper, R.T.   145
Taurus   86

taxation   6, 27-33 *passim*, 39, 107, 119,
    186
tea   17, 33, 44, 236, 242
techniques
    cultivation   3, 9-10, 89, 125-9, 151,
        236-8, 242, 266, 294n.48
    irrigation   3, 79, 80, 82, 94 *see also*
        qanats
Tehran   35, 59, 63, 90, 114, 127, 184-6
    passim, 188, 219-20, 245
Tennessee Valley Authority   84, 100
tobacco   17, 33, 85
Torbat-e Haydarieh   15, 128
topography   11-25, 234-5
trade   33, 40, 41, 175, 234, 263 *see also*
    exports; imports
transport   2, 37, 39, 54, 167, 192, 223,
    257, 274n.36
tribalism   6, 14, 27, 30, 34, 44, 58-9, 68,
    145, 219
Tudeh Party   107
Turkey   146, 234, 243
Turkoman areas   205, 219
tuyuls   28, 30

'ulama   35-6, 58, 110, 111, 189, 190, 206,
    207, 221
underemployment   216, 290n.55
unemployment   50, 63, 177, 216
UNESCO   106
United Nations   43, 106
United States   31, 48, 49, 51-5, 99, 107,
    109-11, 135, 154, 158, 191, 228, 253,
    267n.6, 293n.43
    Department of Agriculture   232
    Development and Resources
        Corporation   99, 135
    Operations Mission   49, 106
urbanization   183-4, 188, 191, 194, 220-
    1, 261-2, 265

Valian, Dr   118
vaqf   36, 51, 58 116, 128-9, 144, 189,
    275n.29
Varamin   182, 204
vegetable growing   14, 16, 84, 86, 177,
    236, 247
Village Councils Act (1956)   106
Voshmgir dam   172

wages   9, 177, 216, 227, 258, 294n.52
war
    with Iraq   97, 196, 218, 219, 223, 228,
        249, 261, 262

World, First 30, 31; Second 41-3, 159, 165
water 4, 5, 8, 13, 19-20, 43-4, 50, 65-104, 121, 129-32, 156, 170, 203, 218-19, 224, 235, 237-8
  laws 69-75
  National – Authority 131
  nationalization 50, 72-4, 132, 143
  rates 73, 132
  rights 69-75
  storage 20, 68, 73, 81, 85, 99, 100, 105, 132, 166
welfare 5, 161, 162, 194
wells 22, 70, 74, 78-9, 90, 95, 96, 98, 100-2, 104, 131, 147, 203
Western Kavir basin 76-7

westernization 154, 190
wheat 8, 15, 16, 18, 24, 33, 86-8 *passim*, 223, 230-4 *passim*, 236, 243, 247, 249, 251, 253, 254, 262
'white revolution' 55-7, 62, 72, 143, 189
women, enfranchisement of 57, 59, 143
Wright, Sir Denis 41

Yazd 14, 77, 94, 121, 130, 138, 277n.10

Zabol 80
Zagros 6, 11, 13, 14, 16, 19, 34, 44, 58-9, 79-83, 86, 219, 238
Zali, Abbas Ali 200
Zenjiabad 144, 147